Shingle spits and salt marshes in the Colne Point area of Essex: a geomorphological study

R.J. Butler
J.T. Greensmith
L.W. Wright

Occasional Paper No. 18 June 1981

Department of Geography,
Queen Mary College,
University of London

ISBN 0 904791 18 1

Cover: Aerial photograph of Colne Point, Essex
(Cambridge University Collection, copyright reserved).

THE AUTHORS

R.J. Butler graduated at Queen Mary College in 1974 and gained a Ph.D. in 1978. Since leaving College he has worked firstly with I.B.M., and is currently working in the computing section of the National Foundation for Education Research in England and Wales.

J.T. Greensmith is University Reader in Geology at Queen Mary College. With E.V. Tucker he has co-authored many papers of a geotechnical and sedimentological nature on Holocene coastal deposits. He is author of *Petrology of the Sedimentary Rocks* (Allen and Unwin, 1978) and *Geology for Schools* (Blackie, 1973).

L.W. Wright is Lecturer in Geography at Queen Mary College. He is co-author with P.J. Wanstall of Occasional Paper No. 9, *The Vegetation of Mediterranean France: a Review.*

ACKNOWLEDGEMENTS

One of us (R.J.B.) wishes to acknowledge the financial assistance received from N.E.R.C. We should all like to thank the Cartographic Unit at Queen Mary College for so skillfully drawing the figures, Lesley Crawford for excellent typing, and Helen Parker who undertook the printing with her usual combination of speed and efficiency.

PREFACE

The coastal marshes and shingle-complex in the Colne Point area of Essex have been studied intensively since 1965 by J.T. Greensmith and E.V. Tucker. Two post-graduate studies have been integrated recently into this general programme of research. One, on Hamford Water by G.J. Leeks, is presently being written up. The other, on the Colne Point area, was completed by R.J. Butler in 1978, and forms the basis of this Occasional Paper. Continued research since 1978 has added further to our understanding of the Holocene history of the whole of the Outer Thames area. Where appropriate to the Colne Point area, these findings have been incorporated into this paper. The detailed results relating to the whole of the Outer Thames area will be published subsequently.

ABSTRACT

The Colne Point area of Essex, England, forms part of the extensive Outer Thames wetlands. The study area comprises a shingle ridge of complex form and structure, in the shelter of which is a well developed salt marsh. This paper examines aspects of the geomorphology of both the shingle ridge and the salt marsh.

The shingle ridge consists of a series of old ridges now surrounded by salt marsh deposits and a contemporary ridge. It is suggested that the older ridges relate to former high sea levels, possibly between 1500 and 1200 B.P., and again between 1200 and 250 B.P. The modern shingle ridge is presently breached in two places where minor spits have developed. The growth of these features and the distal end of the main ridge are examined.

In common with many Essex marshes, the saltings behind the shingle ridge show evidence of erosion. The main causes of erosion are examined. The morphological features of the marsh are described, special emphasis being placed upon the development of the salt pans. Finally, the evolutionary history of the area is examined.

CONTENTS

List of Figures

List of Tables

I INTRODUCTION 1

Nomenclature 4

Literature Relating to the Area 5

Techniques 6

II THE SPIT-COMPLEX 7

Fossil Ridges 7

Modern Ridge 8

The evolution of the spit-complex 8

Lee Spit 11

Horsewash Spit 16

Sandy Spit 21

Sediment supply to Colne Point Spit 22

III THE SALT MARSH 23

Mature Marsh 23

Salt Pans 25

Pan shapes 27

Pan area and size 31

Pan depth 31

Firmness of pan floors 33

Algal mats 34

Recolonization 35

Origin and genesis of salt pans 35

Creeks 36

Creeks in plan 36

Creeks in cross section 38

Possible origin of creek benches 38

Creek levées 40

IV	THE EVOLUTION OF ST. OSYTH MARSH	40
	Stratigraphy	40
	History	43
V	CONCLUSION	45
	REFERENCES	47

FIGURES

1	Location of study area	2
2	Ridge development near Colne Point A: Chenier ridge concept (after Price, 1955) B: Continuous gravel basement concept	9
3a-f	Coastal changes near Colne Point 1874-1973	12-13
4	Changes to Lee Spit since 1950	15
5	Beach profiles near Horsewash Creek showing changes between 1948 and 1976	18
6	Changes to Horsewash Spit 1976/1977	20
7	Geology of Sandy Spit	22
8 a-g	Location map of salt pan mapping units and area of channel study, and maps of pan distribution and size	28-29
9	Development of circular pans associated with creeks	36
10 a-e	Creek profiles showing various morphological forms	37

TABLES

I	Advance and Retreat of Lee Spit	16
II	Details of Salt Pan Mapping Units	27
III	Salt Pan Survey at St. Osyth Marsh	30
IV	Bioturbation and Salt Pan Floor Relationships	34
V	General Sediment Characteristics	41
VI	Detailed Sediment Characteristics	42

I INTRODUCTION

British contributions to the literature on salt marsh geomorphology are surprisingly rare in view of the frequency of occurrence and ecological importance of these wetland sites around our coast. The survey of Scolt Head Island initiated in the decade before the Second World War represents a major exception (Steers 1934, 1958). Studies by Marshall (1962) on the Upper Solway marshes, Pethick (1970) on the North Norfolk marshes, and Butler (1978) on the Essex marshes near Colne Point provide more recent examples. Specific aspects of marsh geomorphology have received intermittent attention. Yapp, Johns and Jones (1917) examined the salt pans of the Dovey marshes and proposed what became a widely accepted theory of salt pan formation which has only recently been re-examined by Pethick (1974) and Kesel and Smith (1978). Chapman (1960) has commented upon the pattern of creek channels as seen in plan, but typically this is a minor part of a much wider ecological study.

The main themes of salt marsh research have been of an ecological, botanical or geological nature, and material of geomorphological interest has been included only incidentally. Examples of this include the surveys of the main marsh surface of the Solway marshes undertaken by Dixon et al (1926) as part of a geological reconnaissance, and of the Milford marshes by Dalby (1970) as part of a botanical study. The important research into sedimentation rates at Scolt Head Island undertaken by Steers (1948), is typically but part of a much wider study. Sediment studies by Richards (1934) on the Dovey marshes and Ranwell (1964) on the Spartina marshes of Southern Britain further illustrate this point.

The evolutionary development of marshes, closely linked as it is with the development of a vegetation succession and especially with sea level changes, has a more widespread literature. Once again, the work on Scolt Head Island serves as a landmark (Steers, 1934), but among other studies one can cite those by Robinson (1952, 1953) and Greensmith and Tucker (1975) on the

Essex coastal marshes, Packham and Liddle (1970) on the Cefni marshes of Anglesey, and Pethick (1980) on the North Norfolk marshes.

An emergent theme of these studies has been the uniqueness of individual marshes with respect to their ecological, botanical and geological conditions. This raises the question as to whether each marsh is also morphologically distinct. The present research was initiated in order to provide a detailed geomorphological account of the Colne Point area (figure 1), an hitherto little studied locality, with the additional objective of assembling information of use to geomorphological studies of British salt marshes in general.

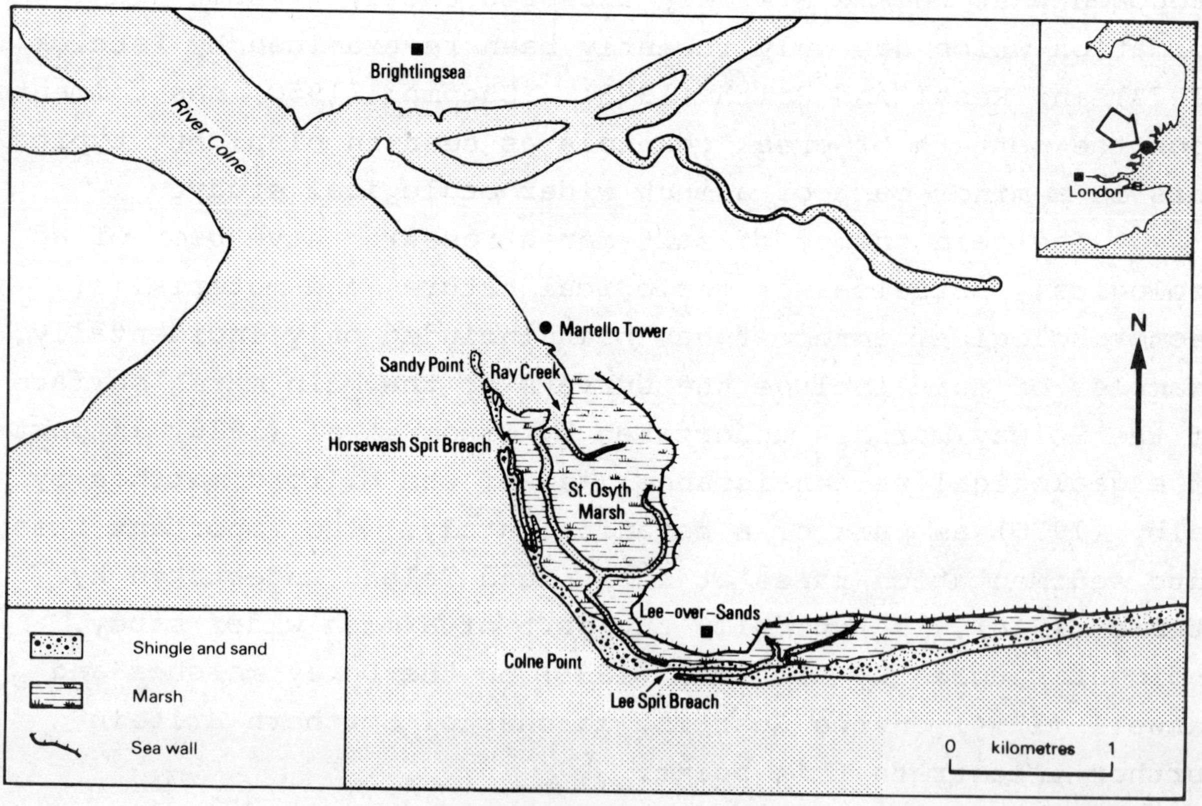

Figure 1: Location of study area.

The marshes at the mouth of the River Colne seven km. west of Clacton-on-Sea are among the most extensive of the Essex wetlands, covering 3.50 sq. km. They consist of mature salt marsh backed by a sea-wall and protected from severe wave attack by a major shingle-complex. Reclaimed land behind the sea-wall still preserves traces of tidal creeks, pointing to the former existence of more extensive marshes.

The history of the saltings is intimately linked with that of the shingle-complex. It is, however, very difficult to set this shingle-complex within a general framework of spit-like features in Britain as has been attempted for the salt marsh. The fact that this structure, except at Sandy Point, its extreme distal end, is today backed by saltings makes its designation as a spit equivocal. It seems inappropriate at this stage to classify it alongside such classic structures as Hurst Castle Spit, Spurn Head and Orfordness, all of which are characterised by a lengthy section free from contact with the land. Unlike these examples and others where growth and attenuation have been especially important geomorphologically over the past century the main interest in the Colne structure is the process of breaching and secondary spit genesis brought about by landwards over-roll across the adjacent marsh surface. This process is particularly well displayed at present in the upper levels of the structure.

Landwards over-rolling is perhaps more closely associated with certain varieties of chenier. The Essex coast has some noteworthy examples of these features, especially at Sales Point and Foulness (Greensmith and Tucker, 1975). Exposure of salt marsh deposits in the inter-tidal zone seawards of the shingle-complex in the Colne Point area, coupled with the obvious landward spread of the shingle, suggests a chenier-like feature. However, it will be shown that elsewhere in the study area the relationship of the shingle structure to the salt marsh is not that typically associated with a chenier. It seems necessary to distinguish two elements within the shingle structure, a basal complex of some considerable age which has a spit-like or bar-like form, and a surficial layer which is being reworked by contemporary processes in the manner of certain cheniers.

Greater understanding of the shingle-complex may be forthcoming when it can be set in its full temporal context. There is much evidence (discussed later) to suggest that this feature dates from at least 4500 B.P. and possibly earlier. At that time, the palaeogeography of the Essex coast in general and the Colne Point area in particular must have been distinct from what exists today. Mean sea level was undoubtedly lower by 3-4 m. (Greensmith and Tucker, 1976). This alone would have been reflected in a different position of the coastline and estuaries, quite apart from the fact that on this low-lying coastal zone small changes in sea level may result in substantial lateral changes in the position of the shoreline. Pleistocene gravel terraces, located at and below 0 m. O.D., which appear to have been the main source of material for the shingle-complex, were doubtless exposed more widely than at present. Until the palaeogeographic history of the area is fully researched, one can but speculate upon the origin of the shingle-complex. The work of Greensmith and Tucker since 1965 has clearly demonstrated the extreme complexity of the Holocene palaeogeography in this area and has gone a long way towards establishing its unique character among the coastlands of Britain. It is in this context that one must view the shingle-complex which protects the salt marsh from erosion.

Nomenclature

In common with many detailed local studies, existing place-names are inadequate to specify all the components recognised in the field. Colne Point *sensu stricto* refers to that point on the coastline where the shingle-complex abruptly changes direction from an east to west trend to a south-north trend. As such it has little utility in the present study. The shingle-complex, of which Colne Point is a part, will be referred to as the Colne Point Spit. This can be divided into three subsections. These are termed Lee Spit, Horsewash Spit and Sandy Spit respectively, running from south to north. The distal end of the Sandy Spit is designated Sandy Point. This is the terminal point of the whole shingle-complex. The marsh behind the shingle is termed the St. Osyth Marsh.

Literature Relating to the Area

The earliest mention of Colne Point *sensu stricto* appears to be in the 1885 Report of the Committee of the British Association on Coastal Erosion. In this Report, erosion was said to be general and widespread along the Essex coast, and specific mention was made of a 1.60 m. high marsh cliff at Colne Point. It is also indicated that there was no shingle at the site, an interesting observation in view of the fact that Colne Point Spit comprises coarse gravel (shingle) with every indication that this has always been the case. Between 1919 and 1939 published material largely related to a number of geological and archaeological excavations, the most noteworthy being those of Hazzledine Warren (1923 and 1932) and Oakley and Leakey (1937). These deal mainly with the existence of a Pleistocene river channel - the Clacton channel - and a Bronze Age occupation surface - the Lyonesse Surface - and their bearing on the present day salt marsh is limited.

Robinson (1952) completed the only significant geomorphological study of the area. This was essentially a cartographic survey based upon comparative studies of Ordnance Survey maps so as to develop a picture of coastal change. Robinson stressed that erosion was active, but also noted that the alignment of the spit-complex had altered several times between 1874 and 1949 in response to breaches through the structure. He also commented on the marked curvature of the spit, which he attributed to northward refraction of material in response to conflicting waves and current between the River Colne and the open sea. Unpublished notes made by Hiller in 1958 indicate that at that period a general lengthening of the spit-complex was taking place and that the structure was being rolled inland over the saltings, exposing marsh layers on the foreshore.

The marsh behind the spit-complex does not appear to have been studied in any great detail. Chapman (1960) comments briefly that the vegetation of the St. Osyth marsh is very similar to that of the Mersea Island marshes where the general succession is from bare mud through a Salicornietum and Asteretum, to a General Salt Marsh and finally a Juncetum. Hider (1975) in a brief historical review of the area comments that there is little evidence of human interference, although he goes on to list six human activities which could have disturbed the ecology: gravel extraction, military fortification, chalet construction, oyster and fish bed digging, sheep grazing and the extraction of certain plants for food such as *Salicornia stricta*. It seems likely that only the first of these is of geomorphological consequence, and it may have been of more importance than Hider suggests.

Techniques

The surface morphology of the shingle-complex and salt marsh was determined by land survey techniques using a Watts Quickset level and by air photography analysis, coupled with detailed field observations. The geometry of the salt pans was obtained by tape measure and field sketching on to squared paper. Details of 410 salt pans were collected. Creek cross sections were determined by pinning a taut tape across the creek and taking readings to the creek floor at set distances along the tape measure. In all, 51 cross sections were measured.

Map and air photograph analysis were also used to determine the evolution of the marsh over the past century. This aspect of the study follows an established tradition in geomorphology exemplified by the work of Robinson (1955) on the harbour entrances of Poole, Christchurch and Pagham, and de Boer and Carr (1969) on Spurn Head and Orfordness. The latter paper details the shortcomings of this type of material, noting especially that the maps and air photographs are not necessarily produced when the feature reaches a turning point in its growth. The sedimentary history of the saltings was examined through bore hole and auger hole evidence. Fifty holes were sunk, augmenting the 22 already logged by the Essex River Board. Sample points were based on a random stratified design, rather than a true random selection because of the inherent difficulties of internal location and access.

A Hiller borer was used to obtain cores. Each core was carefully described in the field and augmented by laboratory analysis using pipette techniques. A standard approach was adopted (Krumbein and Pettijohn, 1938, Carver, 1971). This is fully described elsewhere (Butler, 1978). The fieldwork was carried out between 1975 and 1977.

II THE SPIT-COMPLEX

The spit-complex that today protects St. Osyth Marsh is divisible into two broad categories, a fossil feature and an active feature. Part of the former is now being cannibalized by the latter. The fossil feature is very limited in extent, but still can be clearly subdivided into an inner (older) and an outer (younger) ridge. The modern complex in contrast is an extensive shingle ridge stretching from Seawick caravan site (TM 128126) as far as Sandy Point (TM 089147), a total distance of 4.5km. It is now broken in two places, thereby giving three distinct sections. Each of these sections terminates in a rapidly changing distal point. They are called Lee Spit, Horsewash Spit and Sandy Spit respectively.

Fossil Ridges

The inner-most of the two fossil ridges extends for just over 1km. in an approximate north to south direction from a position about 1km. northwest of Colne Point *sensu stricto*. It rises about 0.50m. above the level of flanking marshes and carries a vegetation dominated by *Suaeda fruticosa* and *Agropyron pungens*. "At various places along its length it is marked by washover deltas and bifurcations, the latter being most pronounced at its northern end. The structure also has a narrow central swale running along its length in which contemporary or subsequent clayey silts were thinly deposited" (Greensmith and Tucker, 1975, p. 261).

Some 20-70m. seawards of this oldest ridge, and separated

from it by a zone of thin salt marsh up to 1.0m. thick, are the remnants of a second and younger ridge. Much of this has now been overridden by the modern ridge which is reworking the gravel constituents into its own structure. Enough remains, however, to indicate that this second *Suaeda*-covered ridge was a complex feature characterised by a number of recurves, suggesting a spit-like development.

Modern Ridge

The modern beach ridge which now protects St. Osyth Marsh is in essence a relatively simple feature in terms of its morphology. Adjacent to the two breaks where Lee Creek and Horsewash Creek flow to the sea the ridge is subject to important short term fluctuations, but throughout most of its 4.5km. length it is morphologically undistinguished. It varies a little in width as measured from mean sea level. It is widest towards the distal end of each spit where it may be between 75 and 90m. across. For much of its length it is much narrower, sometimes being less than 50m. wide. Its height likewise varies from place to place. Near the Lee Creek breach it is over 3.0m. higher than it is close to Sandy Point. At its highest point the crest peaks at 3.9 M.O.D. There is a general decline in height along the spit system from south to north.

The Evolution of the Spit-Complex

The oldest features of the spit-complex are the two fossil ridges. These structures which rise up to 0.50m. above the main marsh surface were termed cheniers by Greensmith and Tucker (1975), who, following Price (1955), define them as shallow-based, perched ridges resting on clay along a marshy seaward facing tidal shore (fig. 2a). The implied origin of these features, therefore, was that they were swept onto the marsh surface during storm wave conditions. It follows from this that although these older ridges may represent the marsh edge (or nearly so) at the time of their formation, the marsh was already in existence. Subsequent work has led Greensmith and Tucker to review this idea and they now prefer to call them marsh edge ridges (Pers. Comm., 1980).

Augering in the course of the present study has suggested that the term chenier is probably a misnomer, since it appears that the ridges are connected at depth, both with each other and with the modern ridge (fig. 2b). The evidence for this assertion is based on both aligned and randomly sited auger-holes taken across the marsh between the older ridges. These show the gravel continues at depth from the present spit as far as Ray Creek. It has not, however, proved possible to determine the base of the gravel. Without this information it is impossible to specify the relationship it bears to the marsh sediments. It is certainly conceivable that the innermost ridge may have formed the protective barrier behind which the main marsh began to develop. While this is possible, it is certainly not necessary, for elsewhere along the Essex coast, as at Dengie and at the eastern end of Mersea Flats similar supratidal sediments have accumulated without any direct protection. Moreover, this and the younger ridge discussed below, could themselves have been formed by the later reworking and moulding of an older and larger structure.

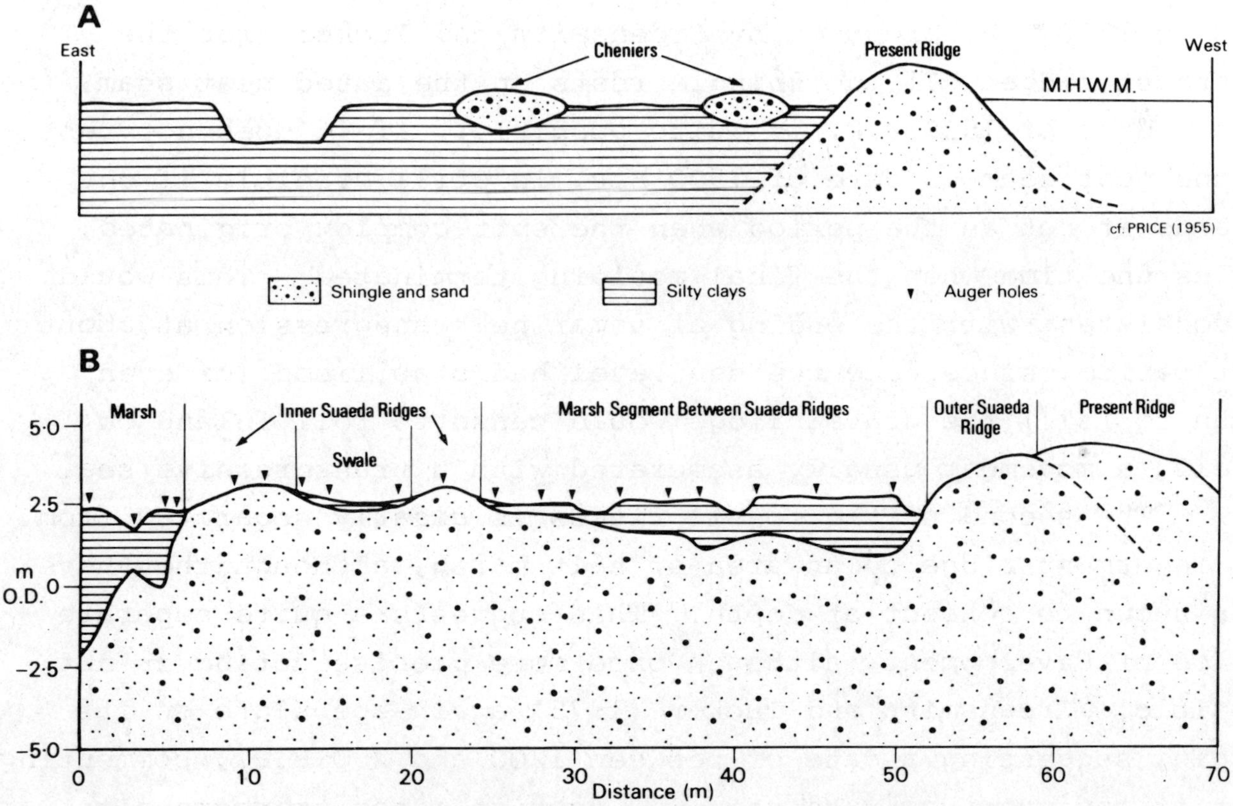

Figure 2: Ridge development near Colne Point.
A: Chenier ridge concept (after Price, 1955).
B: Continuous gravel basement concept.

Unfortunately, direct dating of the innermost (or any other) ridge is not yet possible. Greensmith and Tucker (1975) speculate that the oldest ridge originated between 1500 and 1200 B.P. because there is evidence from the Dengie Peninsula and from the Low Countries that an important marine transgression occurred at that time. A peat sample obtained from within the main marsh during the present study has now been dated radiometrically as 4277± 45 years B.P. This peat occurs within a sequence of silty clays at a depth of about -3.00 metres O.D. These silty clays seem to extend down to a depth of at least -9.00 metres O.D. judging from the evidence supplied in two Essex River Authority (E.R.A.) logs, which are the only ones to have encountered London Clay. The basal silts, although certainly older than the 4277± 45 years B.P. of the peat bed, cannot be dated. Unfortunately, the stratigraphic relationship of the peat to the shingle is still unknown.

The dating of a peat seam landwards of the gravel ridge strongly suggests an even older date for the main ridge itself, but it does not necessarily invalidate the speculative date of 1500 to 1200 B.P. proposed by Greensmith and Tucker for the innermost ridge. If the shingle rests on the dated peat seam, then a date of 1500 B.P. is quite possible. If it does not rest on the peat seam, a date of 1500 B.P. is still credible if one regards it not as the period when the spit-complex originated, but as the time when the final moulding terminated. This would be consistent with the ending of a marine transgression at about that period, since once the sea level had stabilised (or even begun to fall) the gravel ridge would cease to roll inland on itself, a movement usually associated with a transgressive sea.

The second of the fossil ridges is clearly separated from the innermost ridge by an area of salt marsh, although they have been shown to connect at depth. This suggests a quite separate period of development, although once more precise dating is not possible. Greensmith and Tucker (1975) again speculate on its origin, suggesting a date of between 1200 and 250 B.P., commenting that it may eventually be narrowed down to minor transgressive episodes falling between 1000 and 700 B.P., which has been discussed by Bennema (1954) and Tavernier and Moorman (1954).

Once again the difference between this date and the 4277± 45 B.P. of the dated peat seam landward of the ridge is not irreconcilable since the latter may relate to the origin of the structure while the former may relate to a much later remoulding phase.

The evolutionary history of the modern ridge structure is known in more detail, at least for the past one hundred years, but little can be said about its initiation except that it may have evolved from the material of the older fossil ridges when they were more extensive. Cartographic and air photograph analysis provided the basis upon which the evolution of the modern spit-complex is based, but this has been augmented by fieldwork to bring it up to date.

The oldest reliable cartographic evidence traced during this study is a plan of the Nassau estate dating from 1814 which shows the spit quite clearly. It is, however, the successive editions of the O.S. maps, supplemented later by air photographs that provide the most valuable guide to the evolution of the ridge. In the following paragraphs the three subsections of the main ridge are discussed in turn, namely, Lee Spit, Horsewash Spit and Sandy Spit. The main changes are outlined in figure 3a-f.

Lee Spit. On the 1874 Ordnance Survey 1:10560 sheet of the Colne Point area (fig. 3a) a sand and gravel ridge is shown as a continuous feature to Sandy Point. At that time Lee Spit did not exist. The 1897 O.S. 1:2500 sheet likewise indicates that Lee Spit was not present. In contrast, the 1921 O.S. 1:10560 sheet (fig. 3b) shows a marked spit-like feature near Lee-over-Sands.

Two mechanisms for initiating a breach like this suggest themselves. Both involve the over-rolling of the shingle. The first is associated with a major storm surge. The second results from the progressive marshwards roll of the ridge into posterior marsh creeks accomplished quite normally during ordinary spring tides, especially if they are associated with strong winds. This latter mechanism appears to have initiated the Horsewash Creek breach further along the gravel ridge, and has been responsible for breaching a comparable ridge at Sales Point, Essex (Greensmith and Tucker, 1966).

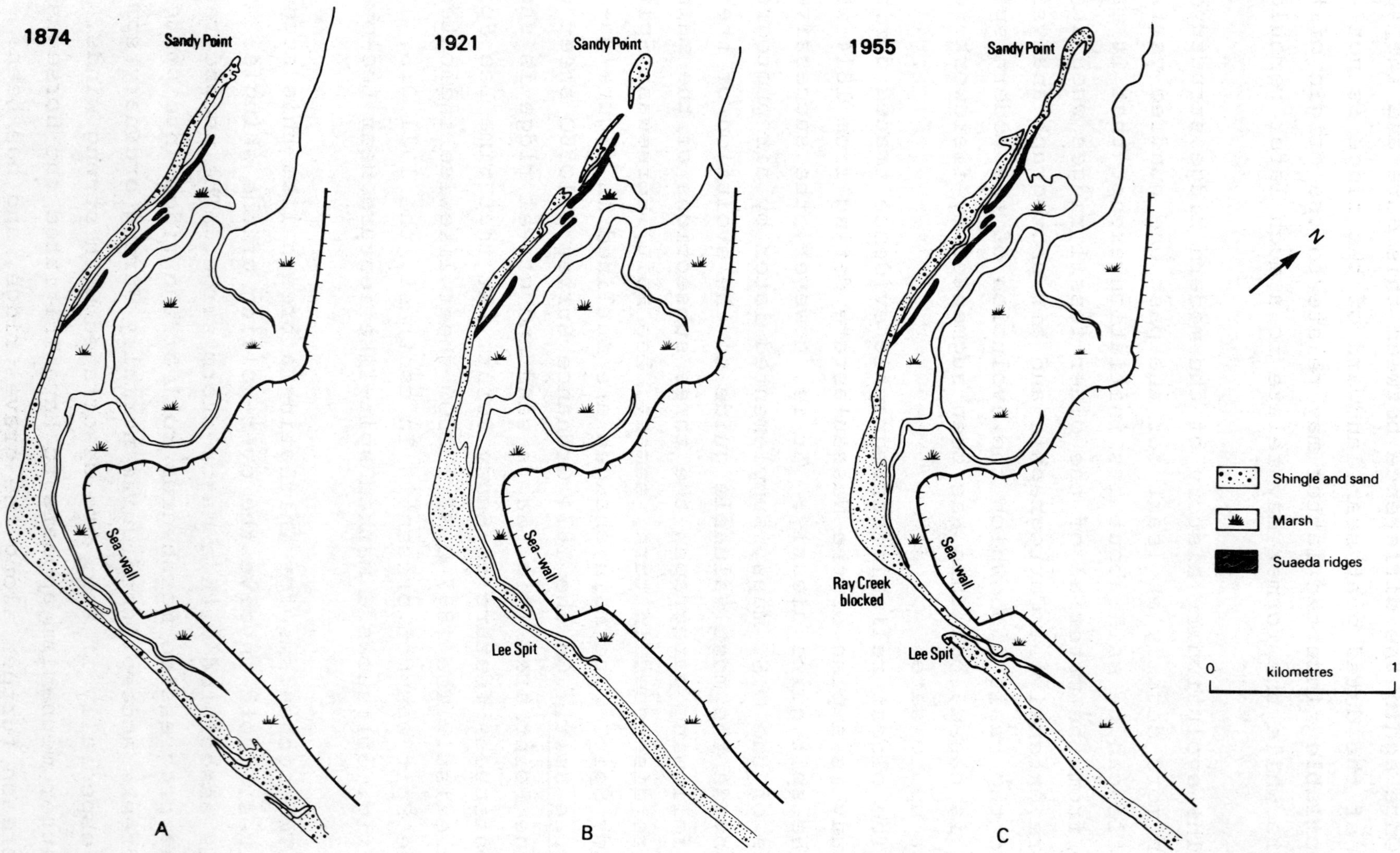

Figure 3: Coastal changes near Colne Point 1874-1973 (A-F).

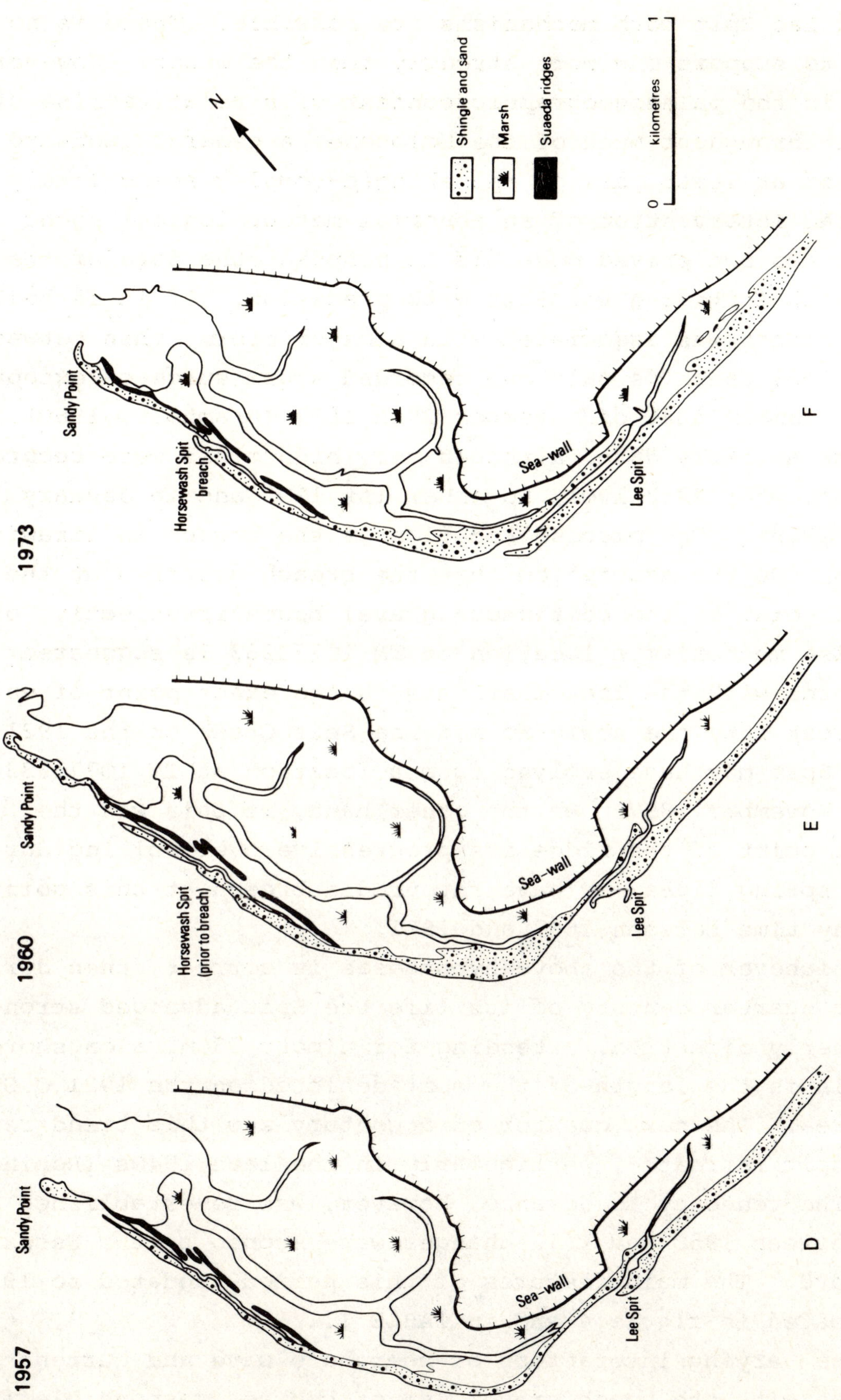

Figure 3: (Continued)

At Lee Spit both mechanisms are possible. There is no evidence to support one more strongly than the other. However, when set in the palaeogeographic context of a relative rise of sea level throughout much of the Holocene, a general landward movement of at least part of the shingle-complex seems likely without the intervention of an abnormal meteorological event.

It has not proved possible to pinpoint the date of the breach which initiated Lee Spit with precision. If it is believed that the breach was associated with a major storm, then between 1897 and 1921 there is only one recorded storm surge of exceptional severity, namely that of November 29th 1897 (Gramholt, 1960). This seems a likely date, although very high tides were recorded also on November 28th 1901, November 1st 1903 and in January 1905 (Grieve, 1959). The precise location of the breach is likewise uncertain. On the assumption that the breach occurred at the narrowest point of the continuous gravel beach irrespective of the precise mechanism a location at TM 10771233 is suggested. This accords with the fact that this is the exact point of entry of Ray Creek into the newly formed Lee Spit Creek on the 1921 map. Thus Lee Spit may have evolved from a location at TM 10771233 on the 29th November 1897. On the other hand, if this was the narrowest point of the ridge its progressive over-rolling during ordinary spring tides may have favoured a breach at this point at almost any time between 1897 and 1921.

Whichever of the above hypotheses is correct, then during the first quarter century of its life Lee Spit advanced strongly in a westerly direction, extending for almost 550m. alongshore, since this is the length of the spit depicted on the 1921 O.S. map of the area. The next quarter of a century saw this trend reversed and the spit retreated, particularly in the late 1940s (Robinson, 1952). The tendency to advance, however, was re-established about 1950. Between 1950 and 1972 changes were mapped by the Essex River Board. The main features of this period, updated to 1976, are indicated in figure 4 and in TABLE I.

The varying interaction of nearshore wave and current conditions with the beach gravel is, of course, responsible for the changes in the position of the distal point, but the effects of man quarrying the gravels may also have had some influence.

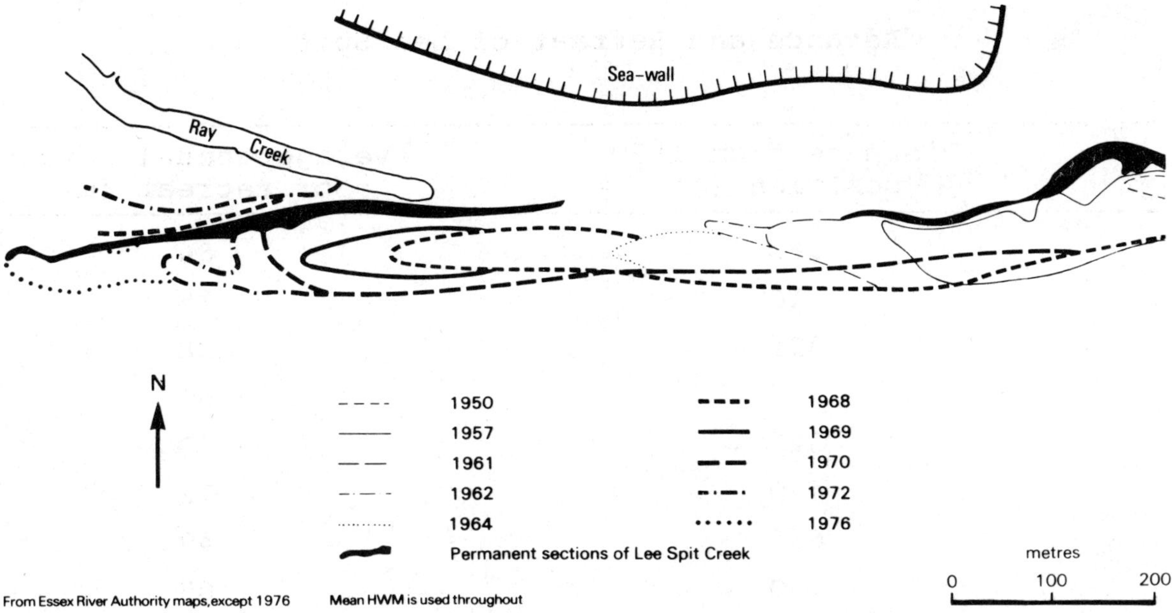

Figure 4: Changes to Lee Spit since 1950.

Indeed, the retreat of the spit during the 1940s and its subsequent regrowth in the 1950s co-incide with active quarrying and the cessation of quarrying respectively.

Concurrently with the advance or retreat of the distal end of Lee Spit since c. 1921 the whole of the spit structure has been eroded on its seaward face so that the position of mean high water has moved landwards. Detailed cross profiling by the Essex River Authority between 1948 and 1956 show this trend. Some of the material eroded from this seaward face has been thrown over the main spine of the spit and deposited on the landward slope, in effect rolling the spit inland. This is shown in the field by the re-emergence of marsh sediments originally deposited to the rear of the spit in the intertidal zone, a particularly good exposure occurring at Seawick (TM 128126)

TABLE I

Advance and Retreat of Lee Spit

Date	Distance from 1950 position (m)	Average annual advance or retreat (m)
1955	180	36
1957	210	15
1960	325	38
1961	375	50
1962	325	- 50
1964	470	72
1968	665	49
1969	730	65
1970	765	35
1972	855	45
1976	1025	42

Horsewash Spit. The near continuous gravel beach that begins at the Lee Creek breach and terminates in the distal point of Horsewash Spit comprises three distinct sections. The first lies behind and extends slightly beyond the position of the distal point of Lee Spit. The second continues about 900m. further along the coast and gives way to the third section where the spit-like form becomes readily apparent.

The first section of beach is notable for the changes that have occurred at the exit point of Ray Creek and the amount of erosion that has taken place further along the coast associated with the position of the distal point of Lee Spit at a site known locally as Henbane Hill. (To avoid confusion, it should be noted that Ray Creek has two exit points, one near Lee Spit, the other through the main marsh. This is not an uncommon condition of salt marsh creeks. It is the Lee Spit exit which is referred to here.) Ray Creek, like Lee Creek, is thought to have gained access to the sea near the present Lee Spit in the great storm of November 1897, though once again marshwards rolling of the ridge could have effected the breach at any time between 1897 and 1921. The 1921 edition of the 1:10560 O.S. map indicates the breach. The 1955 edition of this O.S. map, in contrast, shows the exit to be blocked. This latter condition continued until 1964 when a natural exit was again made through the gravel ridge. This gap is still effective. These changes are shown in figure 3b-e.

It would appear that the history of this exit of Ray Creek into Lee Creek is closely bound up with that of Lee Spit. When the Lee Spit is well developed, as it was in 1921, and as it is today, Ray Creek is capable of maintaining its exit, in spite of its sluggish flow rates because the amount of material moving across the exit is severely limited. When in contrast, Lee Spit terminates well short of the exit, as it did in the 1950s, the exit of Ray Creek would be maintained only if the scouring effects of the creek could clear the greatly increased volume of material moving across the entrance by normal longshore drift processes. Clearly it was not capable of doing this. As the spit elongated itself in the late fifties and early sixties so the amount of longshore material moving across the Ray Creek exit gradually diminished, and the exit was re-established in 1964.

Further along the coast opposite the distal point of Lee Spit, current action and wave refraction have caused significant erosion of the gravel beach. At present erosion is concentrated in the area known as Henbane Hill. Field surveys between February 1975 and May 1977 indicate that 16m. were lost in this period. Most of this took place during storms so that the loss of material has been intermittent. It is evident that the node of erosion is highly localised and intimately related to the position of the end of Lee Spit. As this advances so the node of erosion moves along the coast. Once in the shelter of the spit marked deposition occurs.

Beyond the node of erosion the gravel ridge continues in a nearly straight line for almost 900m. northwards towards Horsewash Spit proper. Under natural conditions this would appear to be a relatively stable section of coast with the dominating feature being the slow rolling inland of the gravel. This section, however, has been affected by the extraction of gravel for aggregate, and repeated cross-profiling has indicated that very marked changes to the beach have occurred. Between 1948 and 1956 when gravel digging took place, the beach repeatedly lost material. After gravel extraction had ceased, restoration gradually took place. Several cross sections taken by the Essex River Authority and updated by present field surveys show this (fig. 5).

The evolution of the distal section of Horsewash Spit bears many similarities to that of Lee Spit. However, unlike Lee Spit which, once formed, has been a persistent, if changing, feature, Horsewash Spit has at times merged back into the general gravel ridge. In other words, the breach which led to the development of the spit has not always managed to maintain itself. Fortunately good air-photograph coverage makes it possible to detail the changes.

Like Lee Spit, Horsewash Spit may have been initiated by the great storm of 1897, or possibly by marshwards rolling of the ridge at about that date. There is certainly a breach indicated on the 1921 1:10560 O.S. map, whereas there is not on the 1897 map. However, by the time of the 1946 air photograph cover the breach had been healed and the gravel beach was once

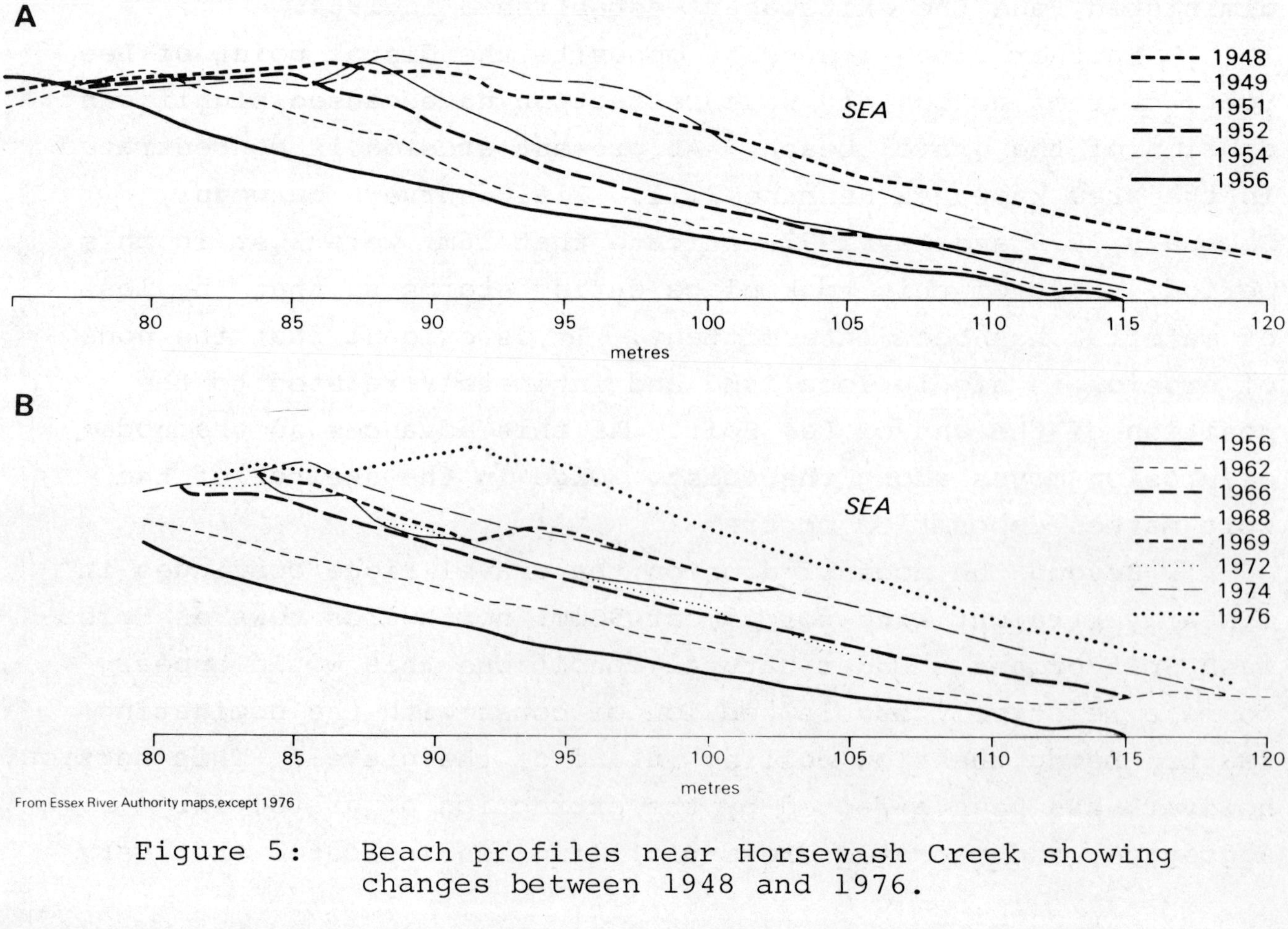

Figure 5: Beach profiles near Horsewash Creek showing changes between 1948 and 1976.

more a continuous feature, although not with quite the same simple alignment as had been previously the case. In fact, repairs to the breach continued through to 1956, by which time the site of the original breakthrough was almost obliterated. Interestingly, the great storm surge of 1953 seems to have had no major effect.

The initial breach occurred at a point where Horsewash Creek swung towards and impinged upon the gravel ridge. Although repaired to a large extent by 1956, this remained the narrowest point on the ridge. Air photograph coverage in June 1956, in June 1957, and yet again in June 1960 all show a narrow but continuous gravel ridge between the creek and the sea. In contrast, a further air photograph survey in August 1961 shows a distinct breach. This can confidently be ascribed to storm and flood action during the winter of 1960-61. Since this date the breach has been maintained, but, as at Lee Spit, the exact point of entry to the sea has varied, although on only a comparatively small scale. Thus in 1976 the distal point of the spit had advanced about 200m. beyond the original breach.

The latest change occurred in the winter of 1976/77 when a new exit point was cut through the spit about 100m. from its distal point. Cut off from its parent body the 100m. of gravel that formed the distal end has been washed across what was the Horsewash channel to join the main gravel ridge. These events are summarised in figure 6.

Concurrent with the evolution of Horsewash Spit recorded above, changes were also taking place in the cross profile of the feature. These were recorded in two Essex River Authority survey lines. The first of these was sited just short of the initial breach point. It therefore exhibits no sudden spectacular change, but in the period 1948-1956 it shows a consistent loss of material from its seaward face and hardly any change to its backslope. The thinning of the ridge in this position due to the frontal erosion is thus clearly indicated. It is suggested that gravel extraction further along the coast at Henbane Hill may have been the reason for the negative sediment balance at this time. When this activity ceased in the mid-fifties,

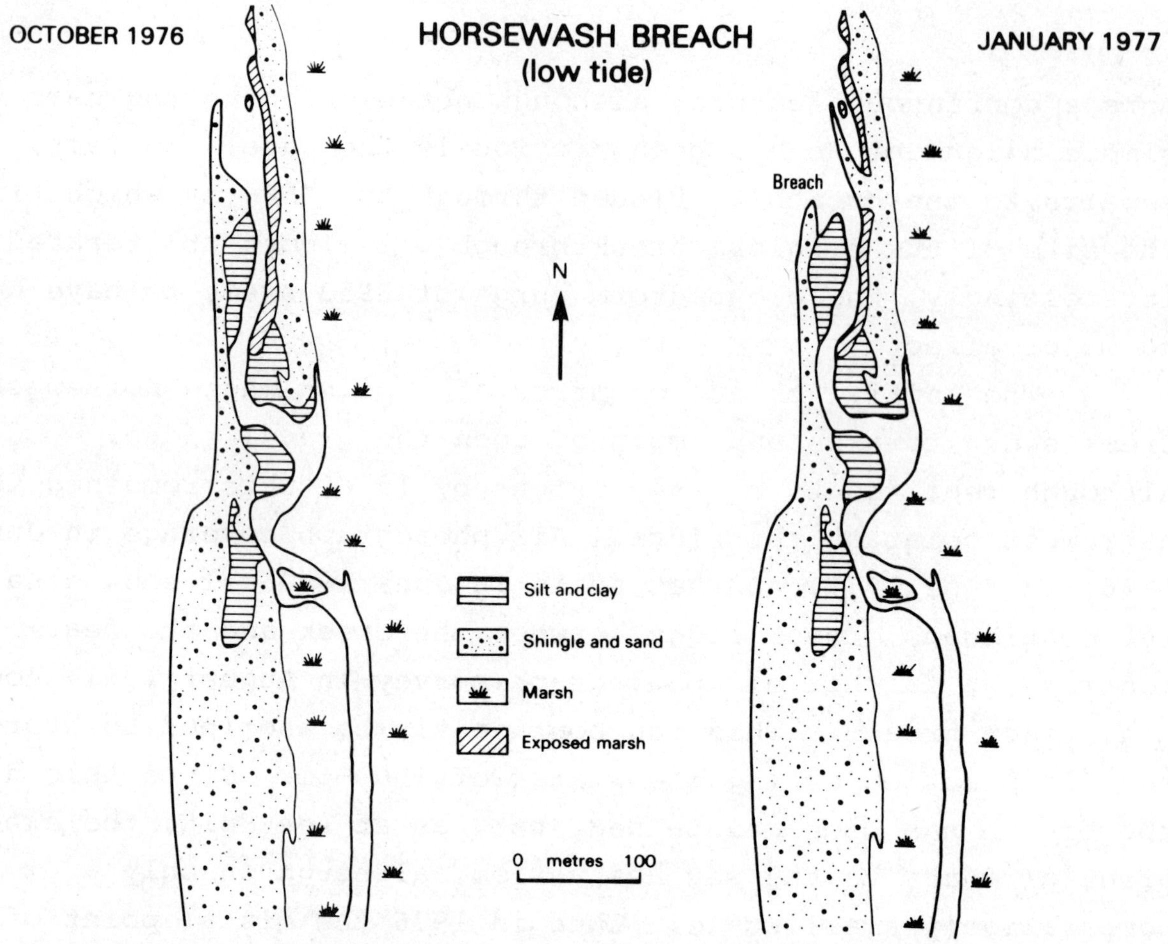

Figure 6: Changes to Horsewash Spit 1976/1977.

conditions were again suitable for a more balanced supply of sediment. A further significant factor may be that it is this section of the Colne Point coastline which has the narrowest intertidal zone. Wave action therefore may be more important here than elsewhere. In fact, after a period of little change lasting until 1966, considerable rebuilding of the beach took place in the next decade.

The second profile was taken just beyond the breach. Thus whereas until 1960 only the gravel ridge is involved, after this date the exit channel of Horsewash Creek and the spit are also included. This makes for a pattern of complex changes. Between 1949 and 1952 slight thinning of the ridge took place. This speeded up in 1953, possibly due to the effect of the February 1st storm surge, but was followed by a period of slow rebuilding until 1956. Conditions then changed, especially following the breach in 1960 when the ridge at this location was starved of sediment and very rapid retreat ensued. In the six years to 1962 the crest of the ridge was washed marshwards almost 40m. compared with only 5m. in the eight years to 1956. By 1974 retreat had continued to the point where the crest was over 80m. landward of its 1948 position. In the two years to 1976 the crest was rolled inland a further 10m. by washover processes inspite of being in the protection of the spit. Horsewash Creek and the new spit appear on the later survey. At first the spit was essentially broad and flat but with a relatively steep face to the creek. By 1976 the height of the spit had increased by about 0.70m. and it had advanced landward by almost 25m.

The fact that the gravel ridge is presently being rolled inland has two consequences. The first is that in places the present gravel ridge is in the process of incorporating the gravels of one of the older fossil ridges within it. So far this involves only the younger of the two *Suaeda* ridges. The second consequence is that marsh sediments are being exposed on the foreshore.

<u>Sandy Spit</u>. Once the breach at Horsewash Spit had been formed, the gravel ridge between that and the extreme distal point of the whole Colne Point complex formed a distinct unit. Sandy Spit differs from both Lee Spit and Horsewash Spit in being far less constrained in its development. Sandy Spit comprises a number of distinct elements. First, there are remnants of the fossil ridges of the spit-complex. As late as 1960, immediately following the Horsewash breach remnants of the younger *Suaeda*-covered ridge could be identified on air photographs with the older *Suaeda* ridge lying landwards of the truncated remains of Horsewash Creek. By 1976 all the younger *Suaeda* ridge had been cannibalised by the present gravel ridge, the creek had been largely obliterated by the over-rolling ridge and the older *Suaeda* ridge now formed an integral part of Sandy Spit. A second element is the modern ridge. This appears to consist of two sections. The first is a sand and gravel ridge that acted as the distal point of the spit until 1960. The other is a new ridge that has developed seawards of the earlier ridge, and extended laterally for almost 250m. since the Horsewash breach in 1960. An accumulation of silty clays occurs between these two parts of the modern ridge (fig. 7).

Beyond the end of the spit is a particularly complex zone of intertidal bars and swales. At low tide a group of these

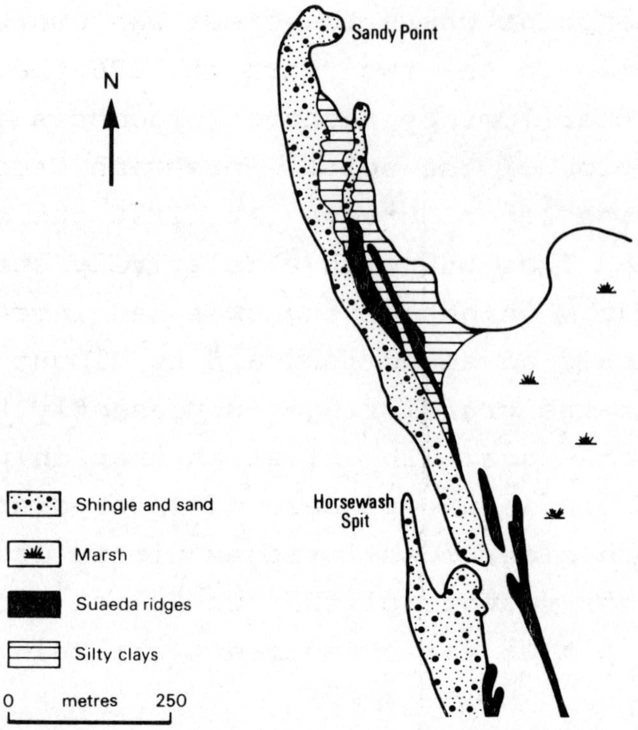

Figure 7: Geology of Sandy Spit.

bars appear to extend the spit by almost 250m.

<u>Sediment Supply to Colne Point Spit</u>. Before leaving the protecting gravel ridges of the spit-complex it is appropriate to discuss briefly the supply of sediment to the Colne Point area. Three possible sources exist. Sediments could be derived from inland via the River Colne, from cliff erosion in the Clacton-Walton area, or from an offshore source. Basically, the spit-complex is a sand and gravel structure and the gravel is predominantly flint. Except on the distal part of Lee Spit, where material less than 8.0mm. in diameter comprises 56 per cent by weight, and on Sandy Spit, the proportion of fines rarely exceed 10 per cent. The coarse fraction is well sorted and tends to range from 14.0mm. up to 20.0mm. along the longest axis.

The nature of the deposit largely precludes the River Colne as a source for the material. Similarly the quantity of material present suggests that local cliff erosion is not the main source. This leaves the offshore zone as the most likely source. This is favoured by Greensmith and Tucker (1975) who point out that large gravel deposits occur in the low-water and mid-water tidal zones along the Blackwater and Colne estuaries and at Mersea Island at about Om. O.D. These banks are interpreted as being reworked terrace deposits of Pleistocene age. The translation of material from these bodies across the intertidal flats is very slow, and marked movement only occurs during prolonged storm conditions.

III THE SALT MARSH

Mature Marsh

Behind the protecting barrier of the gravel ridge lies St. Osyth Marsh. In general terms the marsh is extremely level. Surveyed profiles across the main marsh indicated a mean height of + 2.30m. O.D. $\pm$ 0.15m. and a surface slope as low as 1:625. This uniformity of height means that tidal flooding takes place almost simultaneously over the whole mature marsh. Calculations from Admiralty Tide Tables on the number and duration of tides that rise higher that + 2.30m. O.D., and therefore flood the marsh, yield a figure of 99 floodings per year, which total 34.5 hours or 0.39 per cent of a year. These figures compare with those of 0.41 per cent given by Chapman (1960) for a typical high marsh and 1.90 per cent recorded by Pethick (1970) for the North Norfolk marshes. The low percentage cover in the Colne Point area suggests that sedimentation rates may be very slow.

At present the area of the marsh is being reduced. As has been noted earlier this is well seen where the gravel ridge is being rolled inland and the marsh exposed, and then eroded, on the foreshore. Erosion is also taking place around the main exit point of Ray Creek and on the eastern side of the marsh front. In these areas a low marsh cliff has developed varying between 1.00m. and 1.50m. high. Slumping of blocks from this low cliff is quite common and the fallen material is soon removed. Comparison of the position of the marsh edge in 1874 with that in 1973 suggests a retreat of some 40m. in this 100 year period. Erosion is also taking place in what can best be described as a degraded zone which lies in a shallow gulf behind Sandy Spit at TM 093141. Erosion here is not on a broad front but along the edges of a confused assemblage of low mud runnels. It is not dissimilar to the mud-mound topography described by Greensmith and Tucker (1966) near Sales Point, Essex. The reasons for erosion in this particular spot are unclear, partly because it has not proved possible to determine when it began. There are indications on old photographs that this now bare area was once vegetated, possibly with *Zostera nana* which has declined significantly in the Colne Point area since an outbreak of virus in the 1920s. The loss of this biotic protection may have initiated the present erosion.

Insufficient length of time in the field has prevented the collection of meaningful data on erosion rates. Nevertheless, field observations indicate that erosion processes are perhaps the most important geomorphological events presently operating in the St. Osyth Marsh, a feature it shares with all the marshes of the Outer Thames area. This surely relates to the relative rise in sea level which has affected this area.

Not all the eroded material is lost to the marsh. Some of it is undoubtedly transported up the creek system and, at full tide, spread on to the main marsh surface where some of it is deposited in salt pans or trapped by the vegetation. Layers of coloured sand emplaced on the marsh surface and monitored over a two year period showed variable results. Of fifteen sites on the main marsh, seven were lost through erosion, three had no measurable deposition and five showed a deposition rate ranging from 1mm. to 5mm. per year.

Within the main marsh area the level surface is diversified by salt pans and an intricate creek network. It is to these features that we now turn.

Salt Pans

Salt pans are characteristic features of most salt marsh surfaces. At St. Osyth Marsh, and elsewhere along the Essex coast, salt pans are extremely common. They have been defined by Harshberger (1916) as "..... steep-sided, filled with water by tidal flooding and varying in size from a few feet across to an acre or more in extent", and by Yapp, Johns and Jones (1917) as "..... shallow, basin-like depressions, devoid alike of drainage outlets and of vegetation". Early views on the origin of these pans ascribed them to rafts of decaying vegetation which killed off the underlying salt marsh vegetation and initiated a depression (Warming, 1904, Harshberger, 1916). No mechanism for excavating the depression was proposed. Yapp et al (1917) suggested four ways in which pans may originate, but "in all four methods, the pans originate on bare silt and not on the plant-covered surface of the marsh". The four are outlined below.

(a) Primary pans (or primary depression pans): These were said to form on a primary marsh surface when a series of hummocks coalesced to enclose a shallow depression. Tidal waters persisting in the depression would prevent plant colonisation and leave a circular bare patch as marsh accretion took place around it. Thus the pan was said to originate on the initial surface and to maintain itself throughout marsh growth.

(b) Secondary depression pans: These were formed in exactly the same way, but on a lower, secondary marsh surface, (commonly formed by slumping of the primary surface). They were said to be more numerous and usually deeper owing to the size of fallen hummocks amongst which they originated.

(c) Channel pans: These were said to form by blockages caused by slumping occurring in the channels which led to the stopping of free drainage. Whereas primary and secondary pans were roughly circular in shape, due to tidal eddying, channel pans could be irregular, linear or sinuous depending on the form of the original creek. These pans were thought to occur mostly in the upper part of the channels where vegetation overgrowth was prevalent because of the low frequency of tidal flooding and the ineffectiveness of scour. Often a creek was segmented by the blocking process and a line of pans then was formed clearly indicating the former course of the creek.

(d) Residual pans: this final group, of minor importance, includes small bare depressions that were once part of a much larger pan which had become partly infilled with sediment and vegetation.

It is not entirely clear from the above whether there are strictly four genetically different pans, or just four topographically different forms. Yapp et al (1917) specifically state that pans could not form "..... 'de novo' on existing turf". They also comment that pans once formed tend to persist, but were prone to alteration in shape and depth and tended to become deeper with age. Rather ambiguously, however, they go on to state that "..... as a rule, the higher zones possess fewer pans than the lower". This implies that pan numbers (and density) should decrease with the height and age of the marsh surface, an assertion also made by Warming and Harshberger. This inverse relationship should apply especially to primary pans which could form only on the youthful marsh.

Recent studies have concentrated on the distribution of salt pans within the salt marsh (Pethick, 1970, 1974) and neither process studies of pan development, nor detailed shape studies have been carried beyond the work of Yapp et al. Indeed, Pethick returned to the ideas of Warming and Harshberger as an explanation of his own results.

A complete survey of every salt pan in St. Osyth Marsh was not possible. Furthermore, sampling techniques involving linear surveys or regular random geometrical quadrants presented too many operational problems. It was decided, therefore, to describe all the salt pans that occurred within a given portion of marsh. The main requirements of a selected area were that it should not be too large, that it was basically part of only one morphological unit (all mature marsh or all youthful marsh, etc.) and that it should be clearly defined on all sides, for example by sea walls, major creeks or the spit. In all, seven parts of the marsh were were selected for study, the choice being based on air photographs prior to fieldwork. The seven units are shown in figure 8, and their general features in Tables II and III. The distribution of pans at all seven sites is shown in fig. 8a-g.

TABLE II

Details of Salt Pan Mapping Units

Unit	Grid Ref. of Central Point	Area of Unit (m^2)	No. of Pans
1	TM 093141	22,800	29
2	097144	17,750	71
3	094135	54,050	20
4	096135	13,800	55
5	100135	9,500	95
6	104125	33,300	71
7	111125	14,350	69
			410

Nine characteristics were measured for each individual salt pan out of the total sample of 410 pans at St. Osyth Marsh. These were as follows: pan shape; pan size and area; pan depth; firmness of the pan floor; megafauna content, especially presence or absence of worms and worm casts; floral content, especially presence or absence of algae and enteromorpha; presence or absence of decaying seaweed traces within or adjacent to pan; presence or absence of vegetation on the pan floor; presence or absence of polygonal mudcracking of the pan floor.

When the data had been collected and examined, a number of sub-divisions were possible in certain categories. This gave a total of eighteen measures for the pans of St. Osyth Marsh. These are listed in Table III.

Pan shapes: The most obvious characteristic is probably the most difficult to categorise since at times pans seem infinitely variable both in outline and in size. A subjective estimate of shape into five ordinal classes was made in line with the previous literature: circular, elliptical, linear,

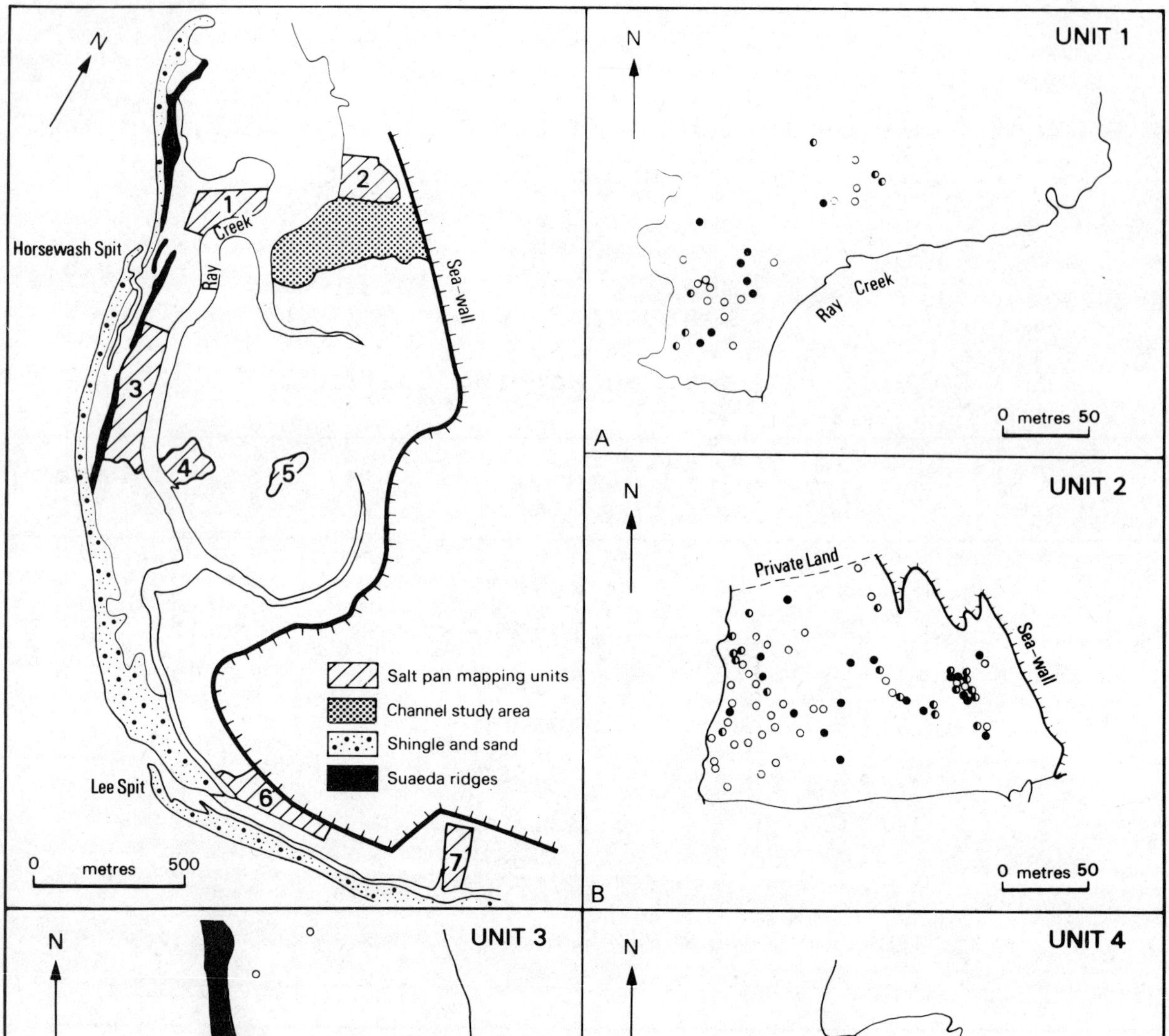

Figure 8: Location map of salt pan mapping units and area of channel study.
A-G maps of pan distribution and size.

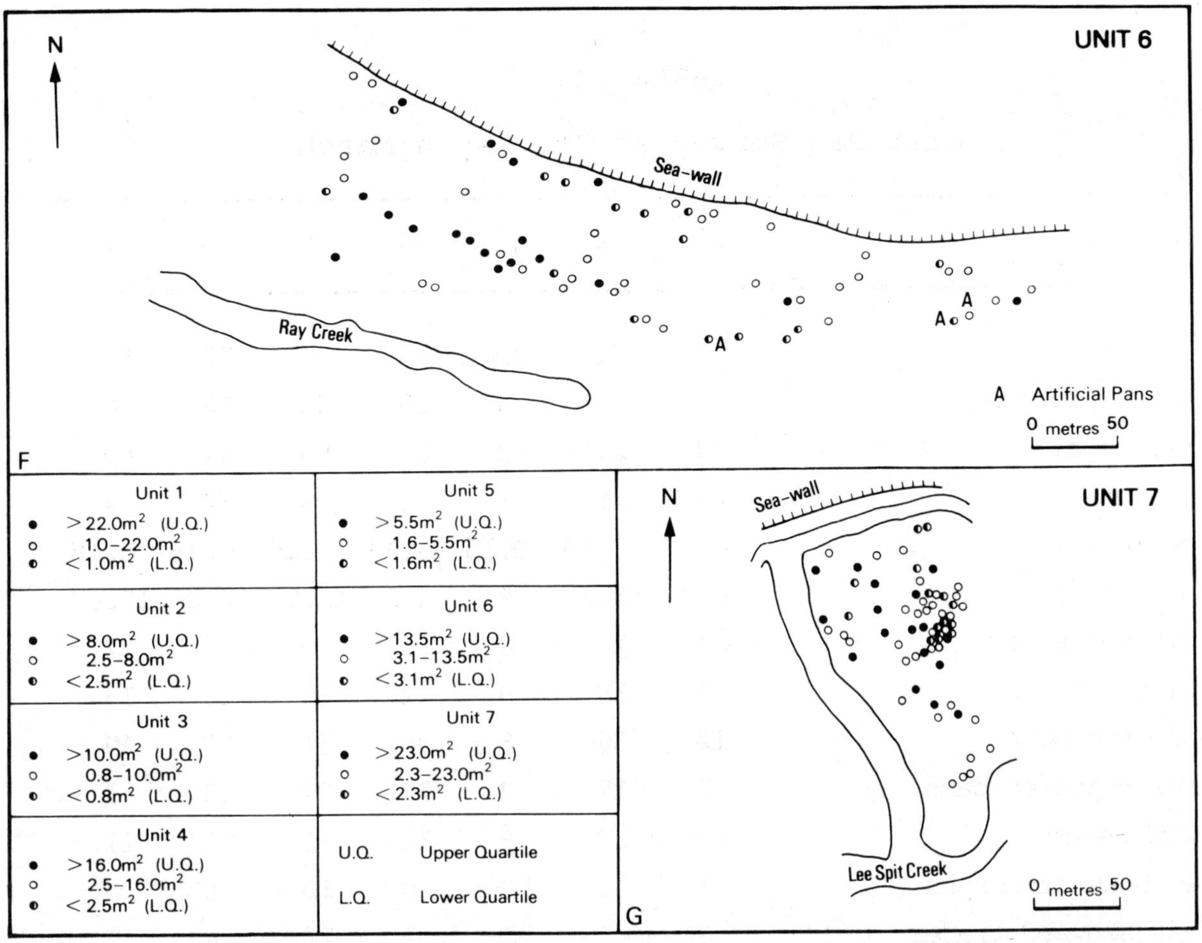

Figure 8: (Continued)

irregular and complex. The results show that 34 per cent of pans were circular or elliptical, 57 per cent linear or irregular, and the remainder complex. If, following Yapp et al (1917), the linear and irregular pans indicate an origin by the channel blocking processes, then this is clearly considerably more important than primary pan formation at St. Osyth Marsh. It may indeed be even more significant than these figures suggest because many circular pans were old circular creek heads and, therefore, intimately associated with channel formation. Against this must be set some, if not all, of the 56 irregular and linear pans which

TABLE III

Salt Pan Survey at St. Osyth Marsh

Unit	1	2	3	4	5	6	7
No. of pans	29	71	20	55	95	71	69
Circular/elliptical	3	39	6	29	30	16	16
Linear/irregular	24	27	13	23	56	41	49
Complex pans	2	5	1	3	9	14	4
Mean depth (m)	0.17	0.18	0.15	0.14	0.15	0.18	0.18
Mean area (m^2)	6.3	15.8	9	4.2	10.3	12.0	15.6
Pans without water	1	17	8	5	0	19	31
Firm-floored pans	11	18	9	2	18	15	19
Soft-floored pans	12	36	8	36	50	33	49
Very soft-floored pans	5	17	3	17	27	23	1
Pans with worms	20	17	8	11	1	23	11
Pans with enteromorpha	1	1	0	0	10	4	0
Adjacent seaweed traces	3	5	0	2	24	4	1
Vegetating pans	5	11	9	6	34	25	3
Polygonal mud cracks	6	25	2	0	1	15	20
Artificial pans	1	4	1	0	0	13	0
Pans with brown algal mats	6	0	0	20	5	7	9
Pans with black algal mats	0	4	1	7	66	10	0

occur in unit five, an area with no contemporary creeks and no indications that any have ever existed. This throws some doubt on the view that all pans of this shape are due to channel blocking.

The relationship between pan shape and the general characteristics of the marsh shows one point of particular note, namely that only three circular pans occur in unit 1 at the front edge of the marsh. This fails to support another contention of Yapp et al (1917) who suggest that primary (and hence circular) pans should be most numerous on the youngest part of the marsh. Some support from another part of their theory is forthcoming from the very mature marsh units (6 and 7), both of which have a low number of circular pans.

There is no clear pattern to the distribution of irregular and linear pans which have high frequencies on all the units.

Pan area and size: Area (size) like pan shape is a surprisingly difficult measure to obtain, particularly once pans exceed 15-20m^2. Moreover, there is nothing in the literature to indicate what significance, if any, size has. This should be borne in mind when examining the quartile area distribution maps shown in figure 8a-g.

The general impression is one of a random spread of both the larger and smaller pans. The one tendency which should be noted is the occurence of larger pans in areas of low creek drainage density, as for example, through the centre of the large, wet, undrained area of unit 5. Smaller pans tend to cluster near creeks or around the edge of undrained ground. There is certainly visual support for this in unit 6.

Pan depth: The depth of salt pans seems to be a crucial parameter with Yapp et al, suggesting an increase in depth with marsh age, and Nichol (1935) favouring the opposite view. In the present study pan depth was measured at intervals along a rigid staff placed across the pan and resting on the sediments of the marsh surface to either side. Three to eight readings were taken in each pan and then averaged. In most cases there was so little slope and so little vertical variation in the pan floor that the mean reading was accurate to $\pm$ 10mm. Not surprisingly, the larger pans showed greater variation.

Pan depth at St. Osyth Marsh ranges from 0.01m. to 0.60m. The lower figures occur in small bare patches particularly in wet, poorly drained areas, while the high upper figures occur in pans developed in secondary marsh. A more typical range is from 0.05 - 0.40m. The overall mean depth is 0.17m. Reference to Table III indicates the constancy of pan depth across all seven units. This suggests that pan depth relates to some factor which is operative all over the marsh. Scour by tidal currents is thought to be the most likely one. It is suggested that the pan depth reflects a critical level below which entrainment of particles by low current velocities cannot take place. Any new sediment laid on the floor will raise the level and thereby permit renewed erosion. Thus over time an equilibrium state is established with scour balancing sedimentation.

If this is the case, then if the character of the pan floor changes so the depth of erosion might also change. Most of the salt pans have soft floors of wet sediment. These are taken as the norm. Two forms, however, differ; one is characterised by a firm floor, the other by a floor covered in part at least with an algal mat. Following Halliwell and O'Connor (1967) it is suggested that if tidal flow crosses a more resistant surface, then scour is reduced. This leads to the supposition that firm-floored pans should be shallower than soft-floored pans (firm-floored pans are defined as those that will support the weight of a man). Ninety-two pans are firm-floored and a Student's t-test was employed to compare the difference between their mean depth and the mean depth of all the other pans in the survey. It was found that with mean depths of 0.14m. for firm-floored pans (σ = 0.05m.) and 0.17m. for all others (σ = 0.06m.) the standard error of the difference was 0.01. This gave a t-value of 4.21 which with 408 degrees of freedom showed a significant difference between the two data sets at more than the 99.9 per cent confidence level.

The second factor possibly inhibiting tidal scour is the presence of an algal mat on the floor of a pan. These mats of living algae are up to 10-20mm. thick and must protect the loose bottom sediments from erosion. The depths of 135 pans with algal mats were compared with 275 pans without, again using the Student's t-test. Pans with mats had a mean depth of 0.15m. (σ = 0.06m) and those without 0.18m. (σ = 0.06m.). This gave a standard error of the difference of 0.01 and a t-value of 4.46 which with the same degrees of freedom, is again significantly different at more than the 99.9 per cent confidence limit.

Three other factors were noted as having an effect on pan depth, but they affect only a very small number of pans. First, wave splash over the marsh cliff caused affected pans to be slightly deeper; second, pans developed between slumped marsh blocks tended to be very deep (in excess of 0.50m.); third, it is possible that increasing pan area could be associated with an increase in pan depth, possibly through greater turbulence.

Finally, a comparison was made between the depth of channel pans and non-channel pans since Yapp <u>et al</u> (1917) had suggested that the former were quite shallow. At St. Osyth Marsh 145 channel pans were tested against 265 other pans again using the Student's t-test. A value of 2.71 was determined for t. The mean depth of channel pans was 0.18m. (σ = 0.06m.) and that of the remainder 0.16m. (σ = 0.06m.). With 408 degrees of freedom the two sample means are significantly different at the 99 per cent confidence level. This finding again contrasts with the views of Yapp et al (1917) who considered channel pans to be shallower than primary pans.

Firmness of pan floors: This topic has already been mentioned in connection with the pan depth. It is also of significance since in an important paper in 1938 Chapman classified pans into hard and soft bottomed types, and attributed the extreme looseness of sediments to the biotic activity of crabs and worms. This needed to be tested in the St. Osyth Marsh area. In practice, three floor types were recognised, firm, soft and very soft. Firm-floored pans show a tendency to occur either above or actually upon gravel substrates where the greater permeability allows free drainage of the silty clays, and at the landward edge of the marsh where infrequent flooding leads to desiccation and over-consolidation. Soft-floored pans show a strong preference for areas where the immediate foundation comprises several meters of relatively impermeable, less well-drained sillty clays.

This pattern is exhibited to some extent in all seven units. Unit 2 has firm-floored pans near the sea-wall and soft-floored pans near the marsh cliff. In contrast unit 4 in the central marsh area has only two firm pans. Units 3 and 7 on the most mature part of the marsh have three firm- and one soft-floored pans respectively. The effect of a gravel substrate is best seen in unit 6 where the near-surface gravel zone is closely associated with firm-floored pans and the adjacent thick, silty clay with soft-floored pans. Unit 5, a wet undrained area, does not follow the general pattern, having 18 firm-floored pans out of a total of 95. These do not relate to a known distribution of gravel, but may reflect the slightly greater elevation of this unit and the marked shallowness of the pans. These two conditions would combine to promote desiccation.

A second feature of note at St. Osyth Marsh is the association of soft-floored pans and vermiform traces, probably produced by *Nereis diversicolors*. This observation supports Chapman's ideas on bioturbation. Table IV shows the number of pans in each unit which have both very soft floors and vermiform traces.

TABLE IV

Bioturbation and Salt Pan Floor Relationships

Unit	Total No. of pans	No. of very soft pans	No. with vermiform traces	No. of very soft pans with vermiform traces
1	29	5	20	5
2	71	19	17	2
3	20	3	8	3
4	55	17	11	2
5	95	27	1	1
6	71	23	23	13
7	69	1	11	0

The association is by no means complete, but the link is inescapable. In contrast to this only four out of 92 firm-floored pans show any signs of faunal activity. On the other hand, the presence of soft-floored pans without vermiform traces may indicate that the worms colonise preferentially those pans rather than actually create the soft floor.

A third tendency is for pans which have evidently been formed by channel blocking processes to have very soft floors, perhaps because the sediment has not yet fully compacted within the creek segment.

<u>Algal mats</u>: Of the 410 pans in the sample units, 135 contain algal mats. These are between 10 and 20mm. thick and can be divided according to type into brown and black forms. There is no apparent morphological difference between the two. Brown mats occur in clear, well-oxygenated water, whereas the black mats are associated with more turbid and anaerobic pans, particularly in unit 5.

Recolonisation: Persistence of salt pans was regarded as an important feature by Yapp et al (1917). Observations showed that at St. Osyth Marsh at least some pans were being recolonised by vegetation. This implies that if the number of pans on the mature marsh is to be maintained, new pans must be created to replace those that are infilled. Almost 20 per cent of pans in the sample survey showed some evidence of recolonisation but in the case of some of these, especially those very shallow pans not encircled by marsh, vegetation within the pan could be the last vestiges of a once complete cover.

Origin and genesis of salt pans: The above discussion has indicated that existing theories of salt pan formation are inadequate to explain the diversity of features found at St. Osyth Marsh, although the methods of channel pan formation suggested by Yapp et al (1917) appear correct. It is the elliptical and circular pans that pose the difficulties, for while some can be assigned to one or other of the two established processes - the coalescence of hummocks on the primary marsh surface or formation beneath a raft of dead or decaying vegetation - others simply do not accord with these ideas.

A feature of both St. Osyth Marsh and nearby Hamford Water is that many creeks and their tributaries terminate in a near circular head. Evidently if these circular sections could be isolated from the rest of the channel a circular pan would develop. Slumping and more especially blocking by the overgrowing of vegetation are two mechanisms by which this could be achieved. In fact the problem is not to find a process to isolate the circular head, it is to find a process that will initiate it. The work at St. Osyth Marsh, together with detailed studies on Hamford Water by Leeks (Pers. Comm.) suggests that turbulent eddies associated with the flood tide may enter small creeks and cause first, a localised widening of the creek, and later, a rounding by scour processes. The hydrodynamics of this concept remain to be investigated, but every stage is well represented at St. Osyth Marsh. The envisaged sequence is shown diagrammatically in figure 9.

Creeks

The channels or creeks within St. Osyth Marsh are arguably the most striking of the morphological features, yet attempts to described them encounter formidable problems. In this brief summary only some of the most obvious features can be outlined. The interested reader is referred to the fuller discussion in Butler (1978). Two aspects of the creeks deserve attention: first, their form in plan view, and second, their cross-sectional shape.

Creeks in plan: A number of distinct patterns of creek systems has been recognised by earlier workers, including the dendritic form characterised by long branches running more or less parallel and exemplified by the East Anglian marshes, and the intricate and tortuous systems associated with the progressive growth of *Spartina* grasses. Broadly speaking, the pattern of creek development at St. Osyth Marsh lies between these two forms since the drainage network is very complex and dense.

It is now common practice to describe stream systems in terms of a modified Horton analysis (cf. Strahler, 1957). Such

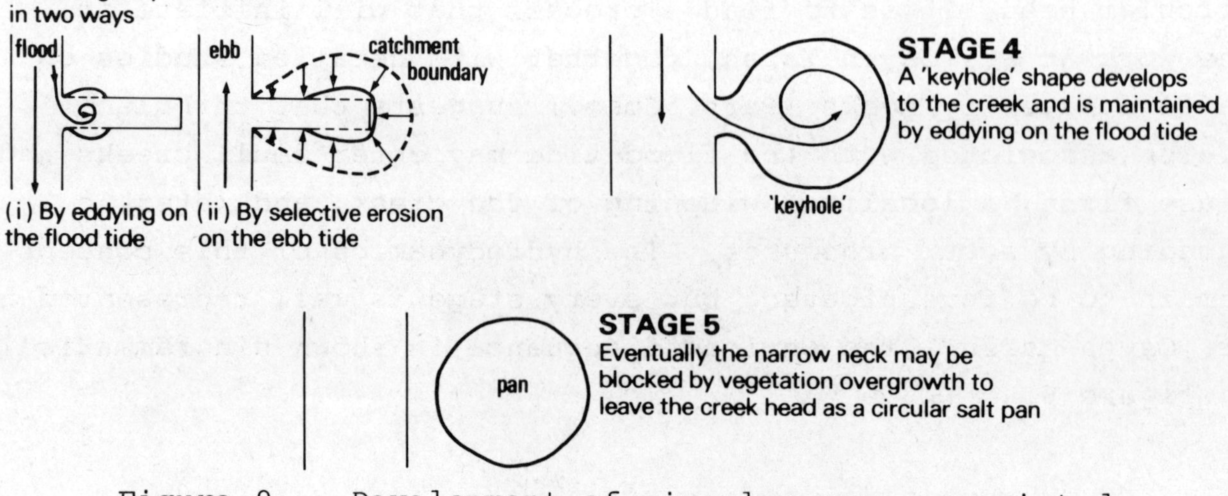

Figure 9: Development of circular pans associated with creeks.

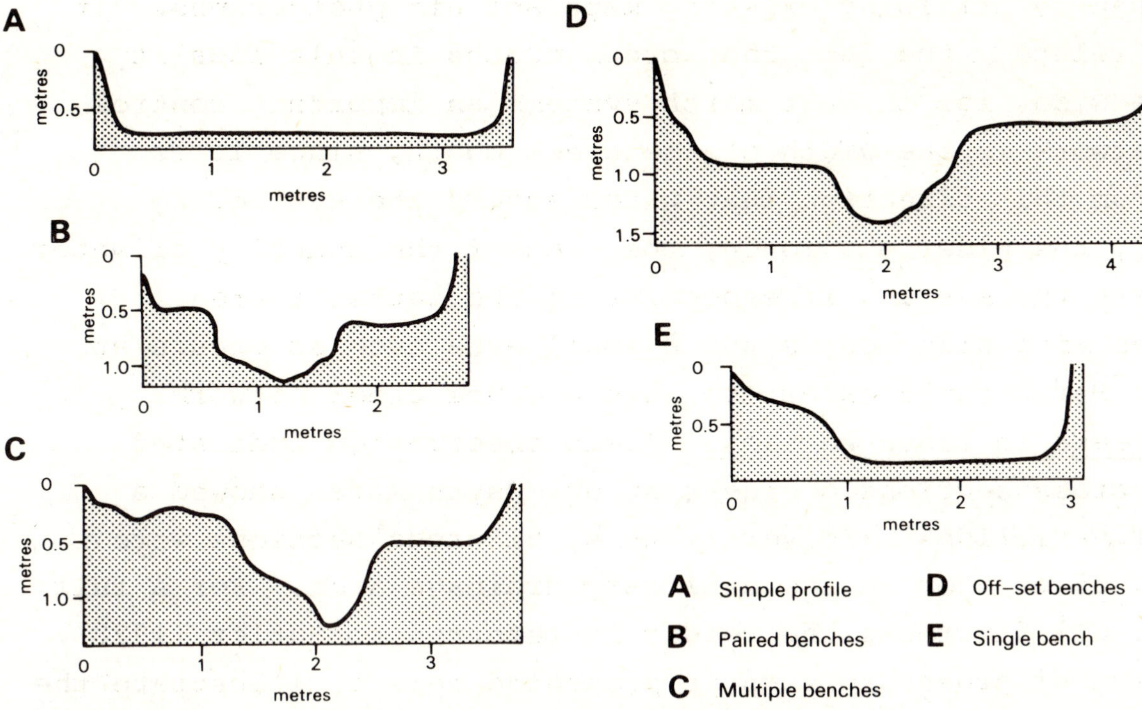

Figure 10 A-E: Creek profiles showing various morphological forms.

an approach is not always easily transferable to a salt marsh environment where additional modes not present in a terrestrial system may be found. It was recognised in an early stage of the investigation that the complexity of the creek network over the whole site demanded that a smaller area of the ground be chosen for more detailed study. At St. Osyth Marsh identification of drainage basins is extremely difficult since many creeks have two or more exits. Thus an arbitrary area was selected largely on the basis of its relatively well defined borders (fig. 8).

Within this selected area the drainage lines were split into first and second order creeks and creeks of higher than second order. There are 443 first order creeks (U) with a total length (L_1) of 5664 metres, and an average length ($\bar{L}$) of 12.8 metres. The total area is 0.26km^2 and the total creek length (all orders) is 17700 metres. This gives a drainage density (D) for all creeks of 6.74. Horton (1945) quotes a figure of 3.00 as the usual maximum for terrestrial systems. The high value at St. Osyth Marsh may in part be the result of the greater accuracy in obtaining absolute creek measurements due to fieldwork methods rather than by utilising existing maps and air photographs. It may also relate to the fact that creek mouths in this locality are quite wide, for in salt marsh systems an important control of the network is the width of the creek mouth, since it is mainly from this direction that water enters the system. Basically, the wider the mouth, the greater the quantity of water that enters the system, irrespective of the latter's area. Localities with wide mouths and a small area such as occurs at St. Osyth Marsh could expect to have a dense creek network.

Creeks in cross section: Field observation indicated that the cross sections of creeks at St. Osyth Marsh showed a number of variations. To verify this, 51 cross sections were measured, 35 of them in the arbitrary drainage area. For a full record of all 51 creeks the reader is referred to Butler (1978). A selection of cross profiles is presented here to illustrate the points made below (fig. 10a-e). The most striking feature of the profiles is the variability, but a number of general points may be made. The first is the tendency of creeks to widen rapidly towards their mouth. One creek increased in width from 1.90m. to 6.30m. over a distance of 300m. The second is the habit of creeks to broaden into a near-circular head. The third point is the general uniformity of channel depth. In the sample survey creek depth ranged from 0.37m. to 1.81m.

The most noticeable cross sectional contrast is between those creeks characterised by a simple profile (fig. 10a) and those that exhibit an intra-channel bench. Within the benched group, four sub-groups can be identified (a) those with a single paired bench of about the same height and extent (fig. 10b); (b) those with two or three benches on one or both sides of the creek (fig. 10c); (c) those with benches occurring at different heights on opposite sides of the creek (fig. 10d); and (d) those creeks which have a bench developed on one side only (fig. 10e). A feature of all four types of benched creek profile is their occurrence near the mouth of a creek.

Possible origin of creek benches: At least four origins may be suggested for bench formation in St. Osyth Marsh. They may act singly but more usually two or more may combine. The four possibilities are as follows.

(a) Depositional origin. This relates to the slumping of creek banks, caused by undercutting and collapse. Benches formed in this manner are usually easily identifiable. In most creeks there are occasional slumped blocks of marsh, but a bench is created only when a long section of bank collapses. The bench usually maintains the original vegetation, often *Halimione portulacoides*, and the rotational movement of the slump leads to the bench surface being angled inwards towards the creek bank. Once slumping has occurred, flow will be constricted and velocities increased. This may lead to incision and the deepening of the basal part of the profile, or alternatively to the gradual erosion of the bench. This is undoubtedly a process of some significance.
(b) Incision of dominant current flow. Pestrong (1965), working on the marshes of San Francisco, ascribes creek benches in that area to erosion by flow maxima. It is a feature of marsh creeks that flow direction varies with the ebb and flood tide. The maximum flow occurs either side of the slack high tide period when the level of water is about 0.30m. below the height of any particular high tide. Since at St. Osyth Marsh only about 100 tides per year exceed the bank-full stage, the dominant current will be acting mainly over a zone 0.50m. to 1.00m. below the bank-full level. The mean depth of the upper profiles of benched creeks in this survey is 0.54m. Incision below this level could be by dominant current action.
(c) Incision relating to through flow creeks. This is a variant on (b) above. Creeks with two exits are often benched although the benches are usually absent from the central section. This absence may reflect the reduced rate of flow where two converging water bodies mix.
(d) Incision relating to changes in the tidal prism. Several factors may have led to a gradual increase in the volume of the tidal prism. A relative rise in sea level for which there is much evidence along the Essex coast may be one cause. Other causes include the shortening of creeks by erosion (40m. in the past century), breaching, or gravel, sand or shell fill during spit migration, thereby reducing creek capacity, and the removal of much salt marsh from tidal effects by reclamation. The

combined effect of an increased volume of sea water entering a reduced creek system could lead to increased current flow and hence incision.

Creek levées; One final point concerning the creeks needs to be noted. In many marshes creeks are bounded by levées which form quite distinctive features. In the case of St. Osyth Marsh they are by no means pronounced, perhaps due to the high percentage of clay in the mobile sediments. Nevertheless, they can be recognised as slightly raised areas extending for up to 10m. away from the creek margins. Their morphological significance is often enhanced by *Halimione portulacoides*.

IV THE EVOLUTION OF ST. OSYTH MARSH

Stratigraphy

In an attempt to determine the long term history of St. Osyth Marsh an analysis of the marsh stratigraphy was undertaken. Before our study began the Anglian Water Authority had drilled 22 holes of various depths into the marsh and spit-complex. The deepest of these passed through 12.5m. and 13.5m. of marsh sediments before reaching the bedrock of London Clay. In this study a further 50 holes were completed, but within the shallow limits imposed by the Hiller borer only one hole exceeded 6.50m. Bedrock was not, therefore, encountered.

Augering indicated that the stratigraphy of the marsh is extremely uniform. The surface layer, varying in thickness between 1.00 and 1.50m. is a moderately firm, oxidised brown clay with traces of rootlets. This is underlain by grey or black, wet, silty clays, which contain few plant remains. Locally, thin, peat-rich seams are encountered (0.10 to 0.30m. thick), but a more widespread seam occurs between -3.00 and -3.20m. O.D. It is a sample from this seam that has been dated at 4277$\pm$ 45 B.P.

Field descriptions of the sediments were augmented by

laboratory analyses. However, it should be noted that since the object of the study was basically morphological rather than geological, the number of sample sites (four) and total number of samples taken (twenty-four) was limited. The sites were selected to determine if there was any marked lateral variation in sediment character across the marsh. The first site was in the main body of the marsh at grid reference TM 10090340; the second was located seawards of Ray Creek at TM 09461332; the third was near the old sea-wall on a very mature portion of the marsh at TM 10181251; and the fourth was on Seawick Marsh at TM 11001262.

Table V lists each of the samples and the proportion by weight of the sand, silt and clay fractions.

TABLE V

General Sediment Characteristics

SITE	SAND % by weight	SILT % by weight	CLAY % by weight
Site 1 sample B103 (Butler, 1976)	5.0	43.5	51.5
Site 2 sample B953	5.0	43.0	52.0
Site 3 sample C102	8.0	38.5	58.5
Site 4 sample X70	6.0	47.0	47.0

Table VI shows the mean percentage by weight of the sand, silt and clay fractions for each site and gives a clear indication of the apparent similarity between the mean sediment fractions in the four auger holes. An analysis of variance conducted on the total sediment distribution yielded an 'F'-value of 3.9, which is not significant at the 95% confidence level. Statistically, therefore, the sediments within the four holes do not vary to any significant degree and in view of the spatial spread of the holes across the marsh, this is a strong indication that the marsh sediment is of a very uniform character. A further 'F' test was

TABLE VI

Detailed Sediment Characteristics

Site	Sand % by weight	Silt % by weight	Clay % by weight	Height O.D.	Depth below marsh surface
Site 1, Sample B103, (Butler, 1976)	7.5	38.6	53.6	1.85m.	0.45m.
	4.6	38.8	56.3	1.80	0.50
	10.6	47.8	41.3	1.05	1.25
	3.1	46.6	49.9	0.18	2.12
	5.3	43.8	51.0	-0.45	2.75
	1.7	42.9	55.0	-1.45	3.75
	2.0	45.1	52.5	-2.45	4.75
	48.1	50.1	1.7	-3.20	5.50
	19.5	55.8	24.6	-3.42	5.72
	6.9	56.9	36.1	-3.70	6.00
Site 2, Sample B953	6.8	37.9	55.3	2.05m.	0.00m.
	1.8	37.9	60.2	1.55	0.50
	6.0	51.0	43.1	1.20	0.85
	2.2	50.3	47.3	0.05	2.00
	6.9	39.6	53.3	-0.75	2.80
Site 3, Sample C102	28.3	24.4	47.2	1.92m.	0.50m.
	2.6	37.1	60.2	0.90	1.52
	1.0	45.0	53.9	0.05	2.37
	1.9	45.8	52.2	-1.55	3.97
	5.1	40.8	54.0	-2.90	5.32
Site 4, Sample X70	4.0	46.3	49.5	1.65	0.65
	3.4	46.7	49.8	1.30	1.00
	4.5	47.3	48.1	0.10	2.20
	8.4	45.5	37.2	-0.45	2.75

conducted to investigate the size distribution of the sediments at broadly similar depths within the four holes. An 'F'-value of 3.5 was obtained, again indicating no significant lateral variation.

The significant feature of the grain size distribution is the very high clay content. The small contribution made by the sand fraction which rarely exceeds 5% by weight, is also noteworthy and rather atypical of British marshes. Overall, therefore, the marsh sediment at St. Osyth Marsh is fine grained, and, in addition, highly cohesive. It is thought that these attributes may have important consequences for creek morphology and the network density, and perhaps be a factor behind the limited spread of *Spartina sp.*

History

The critical feature to emerge from a study of the stratigraphy of the marsh was the discovery of the thin peat seams between -3.00 and -3.20m. O.D. and the determination of a radio-carbon date of 4277 $\pm$ 45 years B.P. Since these seams occur more than 6.00m. above the surface of the London Clay, the origins of the marsh must be much older. Unfortunately, organic material was not present in sufficient quantities in the lower parts of the sequence to permit radiometric dating.

The conditions favourable to the development of salt marshes have been summarised by Chapman (1960, p. 13) as follows: "Maritime salt marshes commonly develop under one of three different physiographic conditions on stable or emerging shorelines. They will develop only under comparable conditions on a submerging coast if the rate of sedimentation is greater than the rate of subsidence. The three physiographic conditions are represented by estuaries, the shelter of spits, off-shore bars or islands, and protected bays. Sometimes marshes develop on what appears to be an open coast, but in such cases it will be found either that the open water in front is not so extensive that serious wave action is possible, or else that the sea is so extremely shallow that, again, large destructive waves do not normally arise."

At present, the marsh is evolving in the lee of a protecting spit-complex comprising two older ridges, and the modern ridge. The significance of this spit-complex to the history of the marsh is unclear. At present it is critical, but this may not have been the case always as unprotected marshes do occur elsewhere along the Essex coast. Augering across the fossil ridges, as shown in figure 2, indicates that the shingle-complex as a whole is a major accumulation. Although its base is undetermined and, therefore, its exact relationship with the salt marsh is unclear, it must have been a controlling factor for a substantial period, and acting on the premise that the present is the key to the past, it may well have been the development of a proto-spit that started marsh accumulation.

Once the original ridge was formed the marsh sediments gradually encroached on its landward margin. The next development was the creation of the two *Suaeda* ridges. The relative building of marsh sediments continued, finally overtopping the oldest ridge, and infilling the depression that existed between the ridges. While this was taking place, the shingle-complex was being reworked along much of its length by a combination of littoral drift and over-washing. More recently it has evolved through a complicated series of breaches to give the present day form.

V CONCLUSION

Notwithstanding the fact that the Colne Point area, like the Essex coast in general, is part of a low to medium wave energy environment, this study has indicated that major changes have been and continue to be an important factor of the geomorphology of the area. The most noticeable changes have affected the gravel ridge that protects the salt marsh behind. It has been shown that this ridge has been breached on a number of occasions, and that today two such breaches are active. Associated with these breaches has been the development of Lee Spit and Horsewash Spit. They have both experienced rapid changes in form, notably Lee Spit, which has been extended by over 1000m. in the last 27 years. The distal point of the whole gravel complex at Sandy Point has been just as active. Superimposed upon these changes has been the tendency for the gravel ridge to be rolled inland over the existing marsh and into some older *Suaeda* covered ridges. The area of salt marsh has, therefore, decreased in the past 80 years even without the effects of reclamation by man which have affected the landward margins.

Within the marsh area itself changes have not been so marked. Nevertheless, peat seams dated 4277 B.P. now occur at about -3.00m. O.D. below the land surface, thereby indicating the extent of sedimentation in this relatively short period of time (the average rate for this period is about 1.0mm. per annum). Within the study period sedimentation rates based on coloured sand layers varied from 15.0mm. to 0.75mm/year. Certain areas of the marsh have been shown to be suffering from erosion, notably the marsh cliff bordering Ray Creek, and an area of mud mounds in the lee of Sandy Point. When coupled with the loss of marsh due to the landwards migration of the shingle-compex, erosion is seen to be the dominant process at present. There is, too, strong evidence to suggest that changes due to vegetation overgrowth and to bank slumping are affecting the creek systems which drain the marsh.

This study has attempted to describe the gravel ridge and marshes near Colne Point, Essex. It has emerged from this that our understanding of the processes operating upon salt marshes is far from complete. This is especially true of salt pans. It has shown also that descriptive techniques applicable to true terrestrial channel systems are not always easily transferred to the salt marsh environment. In setting out to answer questions about the Colne Point area, therefore, more fundamental, and as yet unanswered, questions about salt marsh genesis have been raised.

REFERENCES

Bennema, J. (1954) Holocene movements of land and sea level in the coastal area of the Netherlands. *Geologie Mijnb.*, 16, 254-262.

Boer, G. De and Carr, A.P. (1969) Early maps as historical evidence for coastal change. *Geog. J.*, 135, 17-39.

Butler, R.J. (1978) *Salt marsh morphology and the evolution of Colne Point in Essex, England.* Unpublished Ph.D. thesis. Queen Mary College, University of London, 364 pp.

Carver, R.E. (ed.) (1971) *Procedures in Sedimentary petrology.* Wiley, New York and London.

Chapman, V.J. (1938) Studies in salt marsh ecology. Sections 1-111. *J. Ecol.* 26, 144-179.

Chapman, V.J. (1960) *Salt marshes and salt deserts of the world.* L. Hill, London

Dalby, D.K. (1970) The salt marshes of Milford Haven, Pembrokeshire. *Field Studies*, 3, 297-330.

Dixon, E.E. et al (1926) The geology of the Carlisle, Longtown and Silloth districts. *Mem. Geol. Surv. U.K.*

Gramholt, D.W. (1960) *The coastal marshland of Essex between the seventeenth century and the mid-nineteenth century.* Unpublished M.A. thesis, University of London. 98 pp.

Greensmith, J.T. and Tucker, E.V. (1966) Morphology and evolution of inshore shell ridges and mud-mounds on modern intertidal flats Bradwell, Essex. *Proc. Geol. Assoc.* 77, 329-346.

Greensmith, J.T. and Tucker, E.V. (1975) Dynamic structures in the Holocene chenier plain setting of Essex, England. (eds. Hails, J.R. and Carr, A.P.). In: *Nearshore sediment dynamics and sedimentation.* Wiley, London, 251-272.

Greensmith, J.T. and Tucker, E.V. (1976) Major Flandrian transgressive cycles, sedimentation and palaeogeography in the coastal zone of Essex, England. *Geologie Mijnb.*, 55, 131-146.

Grieve, H. (1959) *The great tide.* Essex C.C.

Halliwell, A.E. and O'Connor, B. (1967) Suspended sediment load in a tidal estuary. *10th Conf. Coastal Eng.*, 1, 687-706.

Harshberger, J.W. (1916) The origin and vegetation of salt marsh pools. *Proc. Amer. Phil. Soc.*, 55, 481-484.

Hider, R.C. (1975) The historical background to Colne Point reserve. *Bull. Essex Nat. Trust.* 4, 24-28.

Horton, R.W. (1945) Erosional development of streams and their drainage patterns; hydrophysical approach to quantitative morphology. *Geol. Soc. Amer. Bull.* 56, 275-370.

Kesel, N. and Smith, J. (1978) Tidal creek and pan formation in intertidal salt marshes. *Scot. Geogr. Mag.*, 94, 159-168.

Krumbein, W.C. and Pettijohn, F.J. (1938) *Manual of sedimentary petrology.* New York. Appleton-Century-Croft.

Leeks, G.J.L. (1975) *Morphology and evolution of Hamford Marshes, Essex.* Unpublished B.Sc. thesis, Queen Mary College, University of London. 81 pp.

Marshall, J.R. (1962) The morphology of the Upper Solway salt marshes. *Scot. Geogr. Mag.* 78, 81-99.

Nichol, E.A.T. (1935) The ecology of a salt marsh. *Journ. Mar. Biol. Assoc. U.K.* 20, 203-261.

Oakley, K.P. and Leaky, M. (1937) Report on excavations at Jaywick Sands, Essex (1934) with some observations on the the Clactonian Industry and on the fauna and geological significance of the Clactonian channel. *Proc. Prehist. Soc.* N.S., 3, 217-260.

Packham, J.R. and Liddle, M.J. (1970) The Cefni Marsh, Anglesey, and its recent development. *Field Studies.* 3, 331-356.

Pestrong, G. (1965) The development of drainage patterns on tidal marshes. *Stanford Studies in Geology.* 10, 1-84.

Pethick, J.S. (1970) *Salt marsh morphology.* Unpublished Ph.D. thesis, Cambridge University, 267 pp.

Pethick, J.S. (1974) The distribution of salt pans on tidal marshes. *Jl. Biogeogr.* 1, 57-62.

Pethick, J.S. (1980) Salt marsh initiation during the Holocene transgression: the example of the North Norfolk marshes. *Jl. Biogeogr.* 7, 1-10.

Price, W.A. (1955) Environment and formation of the chenier plain. *Quaternaria.* 2, 75-86.

Ranwell, D.S. (1964) Spartina salt marshes in Southern Britain. Part II. Rate and seasonal pattern of sediment accretion. *J. Ecol.* 52, 79-94.

Redman, J.B. et al (1885) *Report of the British Association for the Advancement of Science.* Aberdeen.

Richards, F.J. (1934) The salt marshes of the Dovey estuary. IV. The rates of vertical accretion, horizontal extension and scarp erosion. *Ann. Bot.* 48, 225-259.

Robinson, A.H.W. (1952) *The Thames estuary: a regional hydrographic study.* Unpublished M.Sc. thesis, University of London. 110 pp.

Robinson, A.H.W. (1953) The changing coastline of Essex. *Essex Nat.* 29, 79-94.

Robinson, A.H.W. (1955) The harbour entrances of Poole, Christchurch and Pagham. *Geog. J.* 121, 33-50.

Steers, J.A. (ed.) (1934) The physiography and evolution of the land, in *Scolt Head Island.* (ed. J.A. Steers), C.U.P., Cambridge.

Steers, J.A. (1948) Twelve years' measurement of accretion on Norfolk salt marshes. *Geol. Mag.* 85, 163-166.

Steers, J.A. (1958) The physiography of salt marshes. *Proc. Salt Marsh Conf.* Marine Inst., University of Georgia, 3-8.

Steers, J.A. (1960) The physiography, in *Scolt Head Island.* (ed. J.A. Steers), Heffers, Cambridge.

Strahler, A.N. (1957) Quantitative analysis of watershed geomorphology. *Trans. Ann. Geophys. Union.* 38, 913-920.

Tavernier, R. and Moorman, F. (1954) Les changements du riveau de la mer dans la plaine maritime flamande pendant l'Holocène. *Geologie Mijnb.*, 16, 201-206.

Warming, L. (1904) Bidrag til Vadernes, Sandenes og Marskens Natwhistorie. *Mem. Acad. Roy. Sci. et Lettres de Denmark.* Copenhague 7mm Series Section D., Sci. 2.

Warren, S.H. (1923) The *Elephas-antiquus* bed of Clacton-on-Sea (Essex) and its flora and fauna. *Q.J. Geol. Soc.* 79, 606-636.

Warren, S.H. (1932) The palaeolithic industries of the Clacton and Dovercourt districts. *Essex Nat.* 24, 1-31.

Yapp, R.H., Johns, D. and Jones, O.T. (1917) The salt marshes of the Dovey estuary. Part II. *J. Ecol.*, 5, 65-103.